Algebraic Combinatorics and Coinvariant Spaces

CMS Treatises in Mathematics
Published by the Canadian Mathematical Society

Traités de mathématiques de la SMC
Publié par la Société mathématique du Canada

Algebraic Combinatorics and Coinvariant Spaces

François Bergeron

Université du Québec à Montréal

Canadian Mathematical Society
Société mathématique du Canada
Ottawa, Ontario

CRC Press
Taylor & Francis Group
Boca Raton London New York

CRC Press is an imprint of the
Taylor & Francis Group, an **informa** business
AN A K PETERS BOOK

CRC Press
Taylor & Francis Group
6000 Broken Sound Parkway NW, Suite 300
Boca Raton, FL 33487-2742

First issued in paperback 2019

© 2009 by Taylor & Francis Group, LLC
CRC Press is an imprint of Taylor & Francis Group, an Informa business

No claim to original U.S. Government works

ISBN-13: 978-1-56881-324-0 (hbk)
ISBN-13: 978-0-367-38546-0 (pbk)

Library of Congress Cataloging-in-Publication Data

Bergeron, Fran‚ cois
  Algebraic Combinatorics and Coinvariant Spaces / Fran‚ cois Bergeron
    p. cm. -- (CMS treatises in mathematics = Traités de math´ematiques de la SMC)
  Includes bibliographical references and index.
  ISBN 978-1-56881-324-0 (alk. paper)
  1. Combinatorial analysis. 2. Algebraic spaces. I. Title.

QA164.B473 2009
511'.6--dc22

                                                                        2009005442

Contents

Introduction

Whenever we encounter the number $n!$ playing a natural role in some mathematical context, it is almost certain that some interesting combinatorial objects are lurking. This is a recurrent theme in this book, which has a mostly algebraic emphasis. Until very recently a conjecture, known simply as the $n!$ *conjecture*, was still open. Although it is now settled, many important questions surrounding this conjecture remain unanswered. The actual statement[1] of the conjecture makes it appear deceptively easy. Simply stated, the dimension of a certain space of n-variable polynomials had to be equal to $n!$. The only known proof of this (found after at least a decade of intense research by many top-level mathematicians) is rather intricate and makes use of algebraic geometry notions that lie beyond the intended scope of this book.

This is but one instance of the many interesting algebraic incarnations of $n!$. If we expand such consideration to families of integers or polynomials closely related to $n!$, then the richness of the algebraic landscape becomes truly impressive. On a tour of this landscape, we could come across notions such as the cohomology rings of flag varieties, the double affine Hecke algebras of Cherednik, Hilbert schemes, inverse systems, Gromov–Witten invariants, and so on. However, it would be a very tall order to present all of these ties in a book that is intended to be short and accessible, and in fact, most of these subjects are not addressed here; rather, the emphasis is on invariant theory and finite group representation theory. This bias will surely hide much of the beauty and unity of the material considered and undoubtedly makes it a bit more mysterious than it should be. When trying to understand a deep mathematical subject, it is often only with hindsight that we finally understand how simple and clear everything should have been right from the start. A clearer and crisper picture using notions of reductive algebraic groups remains for future consideration.

In the last 25 years, there has been a fundamental transformation and expansion of the scope and depth of combinatorics. A good portion of this evolution has given rise to an independent subject that has come to

---

[1]See Section 10.1 for the formulation.

be known as *algebraic combinatorics*, the goal of which is to study various deep interactions between combinatorics, representation theory, algebraic geometry, and other classical subfields of algebra. One of the nice and rich results of these studies is certainly a renewed interest in the combinatorics of symmetric polynomials, or more generally of invariant polynomials for finite groups of matrices. The origin of this trend can at least be traced back to a 1979 seminal paper by Richard Stanley, "Invariants of Finite Groups and their Applications to Combinatorics" [Stanley 79]. This paper helped initiate a period of joint and complementary efforts involving a large group of researchers, including Björner, Lascoux, Garsia, Schützenberger, and Rota, who were major early researchers of combinatorics.

Another natural line of inquiry leads to the study of quotients associated with the inclusions of algebras that are diagrammed in Figure 1. Most of the spaces at the bottom of this diagram will at least be mentioned in our discussion, and each arrow has a significant role. The top part corresponds to a noncommutative analog of the bottom part, but this top part did not find its way into the final version of this book. Still, it is interesting to bear the whole diagram in mind, both as a background map for our exploration and as a map for further work. Many aspects are missing in this picture, most notably the extensions along a fourth dimensional axis parametrized by the choice of the underlying group action.

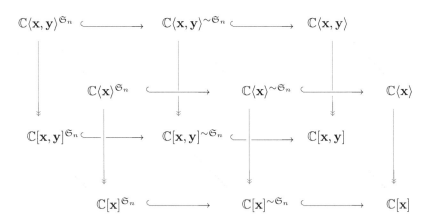

Figure 1. General overview.

The prototypical object in this line is the $\mathfrak{S}_n$-module $\mathbb{C}[\mathbf{x}]/\langle\mathbb{C}[\mathbf{x}]_+^{\mathfrak{S}_n}\rangle$, where $\langle\mathbb{C}[\mathbf{x}]_+^{\mathfrak{S}_n}\rangle$ is the ideal of $\mathbb{C}[\mathbf{x}]$ generated by the constant-term-free symmetric polynomials. This quotient, often encountered in association with the ring $\mathbb{C}[\mathbf{x}]^{\mathfrak{S}_n}$ of $\mathfrak{S}_n$-invariant (symmetric) polynomials, plays an important role at least in the fields of invariant theory [Steinberg 64], Galois theory [Artin 44], and algebraic geometry [Borel 53]. In the first of these contexts, it is known as the "coinvariant space" of the symmetric group; and in the third, it appears as the cohomology ring of the flag variety. It has finite dimension equal to $n!$, and it is in fact isomorphic to the left regular representation of $\mathfrak{S}_n$. This is but the symmetric group case of a more general story concerning finite reflection groups (see [Humphreys 90]). Recent renewed interest in these coinvariant spaces is closely tied with the study of the bigraded $\mathfrak{S}_n$-module of "diagonal coinvariant space" (see [Garsia and Haiman 96a]). In turn these $\mathfrak{S}_n$-modules have been considered in relation to the study of a remarkable family of two-parameter symmetric functions known as *Macdonald functions*. Not only do Macdonald functions unify most of the important families of symmetric functions, but they also play an interesting role in representation theory. It may also be worth mentioning that they appear naturally in the study of Calogero–Sutherland quantum many-body systems in statistical physics (see [Baker and Forrester 97]). On another note, the study of Macdonald functions has prompted the introduction of new families of symmetric functions such as the k-Schur functions (see [Lapointe and Morse 08]), which relate to the study of 3-point Gromov–Witten invariants as well as to WZW conformal field theory.

To keep this presentation short and lively, I have chosen to skip many proofs that can easily be found in the literature, especially when the presentation can easily be followed without explicitly writing down the relevant proof. In these instances, a specific reference is given for the interested reader. However, some proofs have been kept, either because they teach something new about the underlying situation, or simply because they are very nice. One should consider this monograph as a guide to introduce oneself to this subject, rather than as a complete and systematic "exposé" of a more traditional format.

It may be helpful for the reader to consult Appendix A as a guide to notations and formulas.

Acknowledgements

Many people are to be thanked here: some because they have patiently explained to me (in person, at conferences, or in writing) many of the beautiful mathematical notions that are presented in this work, others because they have patiently listened to my sometimes overenthusiastic renditions of these same notions. Happily for me, many are also to be counted in both groups, so that I will not distinguish between them in my thanks. My first thanks go to my close mathematical family, Nantel Bergeron, Adriano Garsia, and Christophe Reutenauer, as well as those (most often friends) who have had such a profound mathematical impact on me: Persi Diaconis, Sergey Fomin, Ira Gessel, Mark Haiman, André Joyal, Alain Lascoux, I. G. Macdonald, Gian-Carlo Rota, Richard Stanley, and Xavier Viennot. Let me add to these the relevant members of our research center "LaCIM",[2] as well as visitors and special friends: Marcelo Aguiar, Jean-Christophe Aval, Srecko Brlek, Frédéric Chapoton, Sara Faridi, Anthony Geramita, Alain Goupil, Jim Haglund, Florent Hivert, Christophe Hohlweg, Gilbert Labelle, Luc Lapointe, Bernard Leclerc, Pierre Leroux (the sadly deceased founder of our research center), Claudia Malvenuto, Jennifer Morse, Frédéric Patras, Bruce Sagan, Franco Saliola, Manfred Schocker, Jean-Yves Thibon, Glenn Tesler, Luc Vinet, Stephanie Van Willigenburg, and Mike Zabrocki. They have all knowingly or unknowingly contributed to this project and, together with all others in the large LaCIM community, they have made all this a very enjoyable daily experience. I also want to thank students and postdoctoral fellows who have been closely tied to the study of the relevant material: Riccardo Biagioli, Anouk Bergeron-Brlek, Philippe Choquette, Sylvie Hamel, Aaron Lauve, François Lamontagne, Peter McNamara, Yannic Vargas, Adolfo Rodriguez, and Mercedes Rosas. They are the ones who may have suffered most from my obsessions, even if they have not yet publicly complained. I hope I have not forgotten anyone else here.

I would like to thank Joel Lewis and Craig Platt for reading through the text and providing suggestions and a list of errata. I am also thankful for the support and encouragement of Jon Borwein and Klaus Peters, who suggested that I write this book, as well as the help of Graham Wright, David Langstroth, and Ellen Ulyanova. I gratefully acknowledge the financial support of the National Science and Engineering Research Council (NSERC) of Canada while I worked on this book.

On a more personal note, let me thank my family for all the morning smiles and their patience while I sat in front of the computer rather than enjoy their company. Merci donc à ma compagne Sylvie et à mon jeune fils Cédrik, ainsi qu'à mes plus grands fils Louis-Philippe et Karl-Frédérik même s'ils vivent maintenant un peu ou beaucoup plus loin.

---

[2] Go to http://www.lacim.uqam.ca if you are anxious to know what the acronym means.

Chapter 1

Basic Combinatorial Objects

Before going on with the more algebraic central theme, it is probably prudent to discuss some general combinatorial background. In this chapter the necessary notions of combinatorial objects will be reviewed in a manner that tries to reconcile conciseness and rigor, while not being too humdrum.

Among the many important combinatorial objects involved in the various interactions between algebra and combinatorics, we should certainly include partitions, tableaux, compositions and permutations. In addition to allowing elementary constructions of irreducible representations of the symmetric group, these notions are involved in the explicit descriptions of many interesting algebraic objects. Moreover, they appear as basic tools for the study of symmetric functions. This is why it is necessary to recall their main properties and consider closely related combinatorial objects.[1]

1.1 Permutations

The symmetric group $\mathfrak{S}_A$, of permutations of a finite set A, plays a crucial role in combinatorics, and this is even more true for algebraic combinatorics. Its manifold appearances range from canonical indexing of fundamental algebraic objects to various natural actions on important spaces.

Let us first review some notations and important concepts concerning permutations of a finite set A. Most often this finite set will be chosen to be of the "standard" form $\{1, 2, \ldots, n\}$. A *permutation* σ of $\{1, 2, \ldots, n\}$ is a bijection $\sigma \colon \{1, 2, \ldots, n\} \overset{\sim}{\longrightarrow} \{1, 2, \ldots, n\}$. Taking into account[2] the usual order on $\{1, 2, \ldots, n\}$ (namely $1 < 2 < \cdots < n$), a permutation can be

---

[1] For more details on such combinatorial objects and their role, the interested reader is encouraged to consult R. Stanley's two volume work [Stanley 97].

[2] This is usually implicit, but we will see in Chapter 6 that there are fundamental reasons to underline that we are actually choosing an order.

described in "one line notation" as the sequence $\sigma_1\sigma_2\cdots\sigma_n$, with $\sigma_i := \sigma(i)$. This presentation makes it evident that there are $n!$ permutations of an n-element set, since there are n possible choices for σ_1, $n-1$ remaining choices for σ_2, etc. We write Id for the *identity* permutation, $\mathrm{Id}(x) = x$. It is the neutral element for the associative operation of composition of permutations, for which we use a multiplicative notation $\sigma\tau$, i.e., $(\sigma\tau)(x) := \sigma(\tau(x))$. Since permutations are bijections, they all have an *inverse* σ^{-1} so that we get a group $\mathfrak{S}_A$ that is called the *symmetric group* of A. We simply write $\mathfrak{S}_n$ when $A = \{1, 2, \ldots, n\}$. An *involution* is a permutation σ such that $\sigma^2 = \mathrm{Id}$.

Exercise. Show that the number I_n, of involutions in $\mathfrak{S}_n$, satisfies the recurrence

$$I_n = \begin{cases} 1 & \text{if } n = 0 \text{ or } n = 1, \\ I_{n-1} + (n-1)I_{n-2} & \text{if } n > 1, \end{cases}$$

the first values being $1, 1, 2, 4, 10, 26, 76, 232, \ldots$.

The *permutation matrix* $M_\sigma := (m_{ij})_{1\leq i,j\leq n}$ associated with σ in $\mathfrak{S}_n$ has entries m_{ij} equal to 1, if $\sigma(i) = j$, and 0 otherwise. The corresponding linear transformation is such that $M_\sigma(x_1, \ldots, x_n) = (x_{\sigma(1)}, \ldots, x_{\sigma(n)})$, and this is compatible with composition: $M_{\sigma\tau} = M_\sigma M_\tau$. Recall that the *order* of a permutation σ is the smallest $k > 0$ such that σ^k is the identity. Its existence follows from the finiteness of the group $\mathfrak{S}_n$. Thus, we have M_σ^k equal to the identity matrix and this implies that $\det(M_\sigma)^k = 1$, which in turn forces[3] the value of $\det(M_\sigma)$ to be ± 1. We can then define the *sign* of a permutation as $\varepsilon(\sigma) := \det(M_\sigma)$. Basic properties of determinants immediately imply that $\varepsilon(\sigma\tau) = \varepsilon(\sigma)\varepsilon(\tau)$, so that the *sign function* $\varepsilon\colon \mathfrak{S}_n \longrightarrow \{-1, +1\}$ is a group homomorphism.

A permutation σ of A is said to be *cyclic* (or called a *cycle*) if for all pairs a, b in A we have $\sigma^i(a) = b$ for some i. It is a well known fact that every permutation decomposes uniquely into disjoint cycles (see Figure 1.1). This is to say that there is a unique partition of A such that the restriction of σ to each block of this partition is cyclic. It is easy to check that two permutations σ and τ are *conjugate*, i.e., $\sigma = \theta^{-1}\tau\theta$ for some θ in $\mathfrak{S}_n$, if and only if they have the same *cycle structure*. This is to say that both permutations have the same number of cycles of length i, for all i. The sign of σ can be computed using this cycle structure as $\varepsilon(\sigma) = (-1)^{n-\gamma(\sigma)}$, where $\gamma(\sigma)$ is the *number of cycles* in the cyclic structure of σ. This gives a direct verification that the sign function is constant on conjugacy classes of $\mathfrak{S}_n$. We denote by $\mathrm{fix}(\sigma)$ the number of *fixed points* of a permutation σ, i.e., the elements k such that $\sigma(k) = k$.

---

[3]Clearly $\det(M_\sigma)$ is an integer.

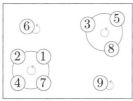

Figure 1.1. Cycle decomposition of 248736159.

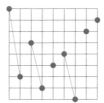

Figure 1.2. The descents of $\sigma = 936245178$.

More Statistics on Permutations

As usual, the *graph* of σ is the subset $\Gamma_\sigma := \{(i, \sigma(i)) \mid 1 \leq i \leq n\}$ of the *combinatorial plane*[4] $\mathbb{N} \times \mathbb{N}$. A value j is said to appear in *position* i in σ, if $\sigma_i = j$. For $1 \leq i \leq n - 1$ we say that we have a *descent* of σ in position i if $\sigma_i > \sigma_{i+1}$. Reading the points of Γ_σ from left to right, a descent corresponds to a position where the next point sits below the previous one (see Figure 1.2). We denote by $\mathrm{Des}(\sigma) := \{i \mid \sigma_i > \sigma_{i+1}\}$ the *descent set* of σ. Let $\mathrm{des}(\sigma)$ stand for the *number of descents* of σ (number of elements of $\mathrm{Des}(\sigma)$). A further important "statistic" is the *major index*

$$\mathrm{maj}(\sigma) := \sum_{i \in \mathrm{Des}(\sigma)} i,$$

defined as the sum of descents. For $\sigma = 936245178$ we have $\mathrm{Des}(\sigma) = \{1, 3, 6\}$, and $\mathrm{maj}(\sigma) = 10$. It is traditional to define the *q-analog of n!* (see Section 1.7) as

$$[n]_q! := \prod_{k=1}^{n} \frac{1 - q^k}{1 - q} \tag{1.1}$$
$$= (1 + q)(1 + q + q^2) \cdots (1 + \cdots + q^{n-1}).$$

It is striking that the *major index generating polynomial* $\sum_{\sigma \in \mathfrak{S}_n} q^{\mathrm{maj}(\sigma)}$ coincides with $[n]_q!$. This is readily verified (recursively) by considering the possible positions for the insertion of n in permutations of $\{1, \ldots, n-1\}$.

---

[4]Throughout this book, $\mathbb{N}$ contains 0.

Also striking is the fact that the *inversion number generating polynomial* is equal to $[n]_q!$. Recall that a pair (i, j), with $i < j$, is said to be an *inversion* of σ if $\sigma(i) > \sigma(j)$. We can then define the *inversion number* (or *length*) $\ell(\sigma)$ of σ as being the number of pairs (i, j) such that $i < j$ and $\sigma(i) > \sigma(j)$. The respective major index and inversion number of the permutation $\sigma = 936245178$ are $\mathrm{maj}(\sigma) = 10$ and $\ell(\sigma) = 16$. The sign of σ can also be computed using the formula $\varepsilon(\sigma) = (-1)^{\ell(\sigma)}$.

Coxeter Generators

To give a "geometric" flavor to inversions we use a *strand presentation* (see Figure 1.3) of permutations. Inversions are then just crossings between pairs of strands. In this representation, permutations are composed by stacking them as illustrated in Figure 1.4. This approach also makes it natural to generalize our discussion to finite reflection groups. More precisely, we reformulate previous definitions in terms of a Coxeter generator-relation presentation of $\mathfrak{S}_n$. Recall that $\mathfrak{S}_n$ is generated by the *adjacent transpositions* $s_i := (i, i+1)$ (see Figure 1.5). These are involutions that satisfy the *braid relations*: $s_i s_j = s_j s_i$ when $|i - j| \geq 2$, and $s_i s_{i+1} s_i = s_{i+1} s_i s_{i+1}$ when $1 \leq i < n$. It is clear that every permutation can be written as a product of such adjacent transpositions. The *reduced expressions* for σ are the minimal length products of this form. In fact, the number of terms in a reduced expression for σ is given by $\ell(\sigma)$. For example, $s_6 s_4 s_5 s_3 s_4 s_3 s_2 s_3$

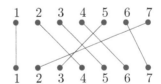

Figure 1.3. Strand presentation of the permutation $\sigma = 1456372$.

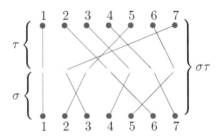

Figure 1.4. Composition of permutations.

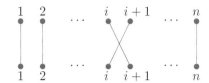

Figure 1.5. Adjacent transposition $s_i = (i, i+1)$.

is one of the several possible reduced expressions (all of length 8) for the permutation $\sigma = 1456372$ of Figure 1.3.

It is now straightforward to translate the notion of descent to this new point of view. Indeed, we check that i is a descent of σ if and only if $\ell(\sigma s_i)$ is smaller than $\ell(\sigma)$. As previously alluded to, the advantage is that we can now generalize the notion of descent to any finite Coxeter group (see Section 3.3).

1.2   Monomials

As a first step for further explorations of the role of the symmetric group in algebraic contexts, let us recall some basic properties of *monomials* in the variables $\mathbf{x} = x_1, x_2, \ldots, x_n$, with the symmetric group acting by permutation of these variables. Extension of this action to polynomials is discussed in Chapter 3.

It is helpful to adopt a *vectorial notation* for monomials. We write $\mathbf{x^a}$ for the monomial $x_1^{a_1} x_2^{a_2} \cdots x_n^{a_n}$, with *exponent vector* $\mathbf{a} = (a_1, a_2, \ldots, a_n)$ in $\mathbb{N}^n$. The degree $\deg(\mathbf{x^a})$ of the monomial $\mathbf{x^a}$ is simply $|\mathbf{a}| = a_1 + a_2 + \cdots + a_n$. Observe that the usual rules for exponentiation hold, so that we have

$$\mathbf{x^0} = 1, \quad \text{and} \quad \mathbf{x^{a+b}} = \mathbf{x^a} \mathbf{x^b}.$$

The *action of* $\mathfrak{S}_n$ on n-variable monomials is defined by $\sigma \cdot \mathbf{x^a} := \mathbf{x}^{\sigma^{-1} \cdot \mathbf{a}}$, where $\tau \cdot (a_1, a_2, \ldots, a_n) = (a_{\tau(1)}, a_{\tau(2)}, \ldots, a_{\tau(n)})$. It is easy to check that the resulting monomial can equivalently be obtained by replacing each x_i by $x_{\sigma(i)}$ in $\mathbf{x^a}$. Observe that there is at least one permutation σ that reorders $\mathbf{a}$ in decreasing order from left to right:

$$a_{\sigma(1)} \geq a_{\sigma(2)} \geq \cdots \geq a_{\sigma(n)},$$

so that the resulting vector characterizes the *orbit* $\mathfrak{S}_n \mathbf{a} := \{\sigma \cdot \mathbf{a} \mid \sigma \in \mathfrak{S}_n\}$ of $\mathbf{a}$ under the action of $\mathfrak{S}_n$. Hence it also characterizes the orbit of $\mathbf{x^a}$. This naturally leads to the notion of "partition", which is described in Section 1.6.

Although monomials are infinite in number, it makes sense to enumerate them degree by degree. It is easy, using a direct recursive argument, to show that there are exactly $\binom{n+d-1}{d}$ degree d monomials in n variables. As will be discussed further in Section 3.1, it follows that this number is the dimension of the homogeneous component of degree d of the space of polynomials in n variables.

Monomials $\mathbf{x^a y^b}$ in two sets of variables (with $\mathbf{y} = y_1, y_2, \ldots, y_n$) are also important for our discussion, especially when we consider the *diagonal action* that permutes the y_k just as the x_k: i.e., $\sigma \cdot y_k = y_{\sigma(k)}$. The orbit of such a bivariate monomial is nicely characterized (from a combinatorial perspective) in the next section. The *bidegree* of $\mathbf{x^a y^b}$ is defined as the pair $\beta(\mathbf{x^a y^b}) := (d, e)$ in $\mathbb{N} \times \mathbb{N}$, when $|\mathbf{a}| = d$ and $|\mathbf{b}| = e$.

1.3   Diagrams

Our next fundamental notion, that of "diagram", allows a classification of orbits of monomials $\mathbf{x^a y^b}$, as well as some bivariate determinants $\Delta_\mathbf{d}(\mathbf{x}, \mathbf{y})$ of importance for our discussion (see Section 10.2). It also comes up under many guises throughout our explorations, in particular as a generalization of the more classical notion of "Young" diagrams (see Section 1.5). We define a *diagram* to be any finite subset of $\mathbb{N} \times \mathbb{N}$. Adopting a geometrical viewpoint, let us call the elements of a diagram *cells*, each cell (i, j) being pictured as a 1×1 square having vertices (i, j), $(i+1, j)$, $(i, j+1)$, and $(i+1, j+1)$. Thus Figure 1.6 depicts the diagram $\{(0,0), (2,0), (2,1), (3,1), (0,2), (0,3), (1,3)\}$. We adopt here the "French convention" and use cartesian coordinates, rather than the "English convention", which uses matrix-like coordinates.[5]

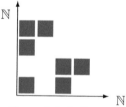

Figure 1.6. A diagram.

For $r > 1$, cells having a second coordinate equal to $r - 1$ are said to have *height* r. The rth *row* of a diagram is the set of its cells that lie at

---

[5]While this convention is becoming more popular in the field, it is contrary to one that is still widely used. We suggest to readers who feel uncomfortable with this that they follow I. G. Macdonald's advice [Macdonald 95, p. 2] "Readers who prefer this convention should read this book upside down in a mirror".

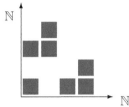

Figure 1.7. Conjugate of the diagram in Figure 1.6.

height r. Likewise, the rth *column* of a diagram is the set of its cells that have first coordinate equal to $r - 1$. The *conjugate* $\mathbf{d}'$ of a diagram $\mathbf{d}$ is the set of cells of form (j, i), with (i, j) a cell of $\mathbf{d}$. Thus, $\mathbf{d}'$ is obtained by flipping $\mathbf{d}$ along the line $x = y$ in $\mathbb{N} \times \mathbb{N}$. This is illustrated in Figure 1.7 for the diagram in Figure 1.6. Observe that conjugation is an involution, i.e., $\mathbf{d}'' = \mathbf{d}$.

One can bijectively encode a diagram as a polynomial in two variables q and t by setting $B_{\mathbf{d}}(q, t) := \sum_{(i,j) \in \mathbf{d}} q^i t^j$, each term corresponding to a cell of $\mathbf{d}$. For the diagram in Figure 1.6 we get

$$B_{\mathbf{d}}(q, t) = 1 + q^2 + q^2 t + q^3 t + t^2 + t^3 + q t^3.$$

We also set $n(\mathbf{d}) := \sum_{(i,j) \in \mathbf{d}} j$ so that

$$|\mathbf{d}| := \sum_{(i,j) \in \mathbf{d}} (i, j) = \big(n(\mathbf{d}'), n(\mathbf{d})\big). \tag{1.2}$$

In later chapters, we denote by $T_{\mathbf{d}} = T_{\mathbf{d}}(q, t)$ the monomial $q^{n(\mathbf{d}')} t^{n(\mathbf{d})}$. Consider now a monomial $\mathbf{x}^{\mathbf{a}} \mathbf{y}^{\mathbf{b}}$, with $\mathbf{a} = (a_1, \dots, a_n)$ and $\mathbf{b} = (b_1, \dots, b_n)$. We write $\mathbf{d} = (\mathbf{a}, \mathbf{b})$ if

$$\mathbf{d} = \{(a_1, b_1), (a_2, b_2), \dots, (a_n, b_n)\}.$$

Observe that the bidegree of $\mathbf{x}^{\mathbf{a}} \mathbf{y}^{\mathbf{b}}$ is $\big(n(\mathbf{d}'), n(\mathbf{d})\big)$, here considered as a multiset. It is clear that the orbit of $\mathbf{x}^{\mathbf{a}} \mathbf{y}^{\mathbf{b}}$ under the diagonal action of $\mathfrak{S}_n$ is precisely characterized by the diagram $\mathbf{d}$. For example,

$$x_1^2 x_2^4 y_2 x_3 y_3, \quad x_1^2 x_3^4 y_3 x_2 y_2, \quad x_2^2 x_1^4 y_1 x_3 y_3,$$
$$x_2^2 x_3^4 y_3 x_1 y_1, \quad x_3^2 x_1^4 y_1 x_2 y_2, \quad x_3^2 x_2^4 y_2 x_1 y_1,$$

is characterized by the diagram $\{(2, 0), (4, 1), (1, 1)\}$. The correspondence between diagrams and orbits of monomials is essential for our later discussion. Given any n-celled diagram $\mathbf{d}$ with some order on its cells, we also define the *lattice determinant*

$$\Delta_{\mathbf{d}}(\mathbf{x}, \mathbf{y}) := \det(x_k^j y_k^i)_{\substack{1 \le k \le n \\ (i,j) \in \mathbf{d}}}.$$

Note that $\Delta_{\mathbf{d}} = \Delta_{\mathbf{d}}(\mathbf{x}, \mathbf{y})$ is bihomogeneous of bidegree $|\mathbf{d}|$. For example, with $\mathbf{d} = \{(0,0), (1,0), (0,1)\}$, we get the bidegree $(1,1)$ polynomial

$$\Delta_{\mathbf{d}} = \det \begin{vmatrix} 1 & x_1 & y_1 \\ 1 & x_2 & y_2 \\ 1 & x_3 & y_3 \end{vmatrix}$$

$$= x_2 y_3 - x_3 y_2 - x_1 y_3 + x_3 y_1 + x_1 y_2 - x_2 y_1.$$

1.4   Partial Orders on $\mathbb{N}^n$

To organize combinatorial manipulations on cells of a diagram, we often need to order them. In fact, we may as well extend our discussion to orders on set of cells in $\mathbb{N}^n$, even if we often restrict our illustrations to 2-cells simply because they are easier to represent. Still, most of the essential features of the general n-dimensional situation are made apparent in the two-dimensional context.

A first interesting order on 2-cells is the *increasing reading order* which corresponds to reading cells from left to right along rows, starting with the top one and going down just as we read a book. Illustrating this with the diagram in Figure 1.6, we order the cells as $(0,3) < (1,3) < (0,2) < (2,1) < (3,1) < (0,0) < (2,0)$. Figure 1.8 shows how to label cells according to this increasing order.

Figure 1.8. Reading order labeling.

In *componentwise order* on $\mathbb{Z}^n$, we have $\mathbf{a} \leq \mathbf{b}$ if and only if $\mathbf{b} - \mathbf{a}$ has all coordinates nonnegative, so that $a_i \leq b_i$ for all $1 \leq i \leq n$. Notice that this is only a *partial order* so that some cells are not comparable. For instance, we have $(1,3) \not\leq (2,2)$ and $(2,2) \not\leq (1,3)$. It is easy to verify that every subset (or multi-subset) E of $\mathbb{N}^n$ has a finite number of minimal elements for this order. The set of these minimal elements is denoted by $\min(E)$. Observe that componentwise inequality between exponent vectors $\mathbf{a}$ corresponds to a divisibility condition between the associated monomials $\mathbf{x}^{\mathbf{a}}$. In particular, any set of monomials affords a finite set of minimal monomials for the divisibility relation. Let us elaborate on this in the $n = 2$ case. The *shadow* of a subset E, of $\mathbb{N} \times \mathbb{N}$, is the set of points (a, b) that lie

Figure 1.9. Shadow of a subset of $\mathbb{N}^2$.

to the *northeast* of some point in E. Reformulated in terms of monomials, (a, b) lies in the shadow of (c, d) exactly when $x^a y^b$ is divisible by $x^c y^d$. The lower left corners of this shadow correspond to the minimal elements of E, as illustrated in Figure 1.9. Clearly, no two minimal points lie in the same column (or same row), thus we can order $\min(E)$ by increasing column value, to get $\min(E) = \{(a_1, b_1), (a_2, b_2), \ldots, (a_k, b_k)\}$, with $a_1 < a_2 < \cdots < a_k$.

Our last order is the total order called the *lexicographic order* on n-cells. For $\mathbf{a}$ and $\mathbf{b}$ in $\mathbb{N}^n$, we say that $\mathbf{a} >_{\text{lex}} \mathbf{b}$ if the leftmost nonzero entry of the vector difference $\mathbf{a} - \mathbf{b}$ is positive. We use the same notation for the associated monomials, writing $\mathbf{x}^{\mathbf{a}} >_{\text{lex}} \mathbf{x}^{\mathbf{b}}$ whenever $\mathbf{a} >_{\text{lex}} \mathbf{b}$.

Monomial Orders

Among the crucial aspects of interactions between combinatorics and algebraic geometry, the "effective algorithmic" emphasis of modern algebraic geometry is certainly a leading feature. This algorithmic approach relies mainly on the possibility of explicit computation of "Gröbner bases" for ideals. We will come back to this but for the moment we concentrate on "monomial orders", an essential ingredient in the computation of Gröbner basis.

The correspondence that we established in Section 1.3 is often used in the actual description of these monomial orders. More precisely a *monomial order* on the set of monomials $\{\mathbf{x}^{\mathbf{a}} \mid \mathbf{a} \in \mathbb{N}^n\}$ (or equivalently on $\mathbb{N}^n$) is a well ordering such that for all $\mathbf{c} \in \mathbb{N}^n$, $\mathbf{x}^{\mathbf{a}+\mathbf{c}} > \mathbf{x}^{\mathbf{b}+\mathbf{c}}$ whenever $\mathbf{x}^{\mathbf{a}} > \mathbf{x}^{\mathbf{b}}$. The lexicographic order is an example of such a monomial order. The verification that the relevant property holds in this case is immediate since $(\mathbf{a} + \mathbf{c}) - (\mathbf{b} + \mathbf{c}) = \mathbf{a} - \mathbf{b}$. Observe that we are implicitly assuming that $x_1 >_{\text{lex}} x_2 >_{\text{lex}} \cdots >_{\text{lex}} x_n$, but we could choose some other order on the variables which would lead to a different monomial order.

A further desirable property of monomial orders is to be *graded*. This is to say that we must have $\mathbf{x}^{\mathbf{a}} > \mathbf{x}^{\mathbf{b}}$, whenever $|\mathbf{a}| > |\mathbf{b}|$, so that monomials of higher degree are greater. For instance, we can turn the lexicographic

order into a graded version by insisting first on a degree comparison. The
resulting order is naturally called the *graded lex order*, and we have

$$\mathbf{x}^\mathbf{a} >_{\text{grlex}} \mathbf{x}^\mathbf{b}, \quad \text{iff} \quad \begin{cases} |\mathbf{a}| > |\mathbf{b}| & \text{or} \\ |\mathbf{a}| = |\mathbf{b}| & \text{and } \mathbf{a} >_{\text{lex}} \mathbf{b}. \end{cases} \tag{1.3}$$

For many computations, the *graded reverse lexicographic order* appears to
be the most efficient (see [Cox et al. 92]). For equal degree monomials this
graded order is obtained by setting $\mathbf{x}^\mathbf{a} >_{\text{grevlex}} \mathbf{x}^\mathbf{b}$ if the rightmost nonzero
entry of $\mathbf{a} - \mathbf{b}$ is negative.

1.5   Young Diagrams

Considering the componentwise partial order on $\mathbb{N} \times \mathbb{N}$, we are naturally
led to the important special notion of *Young diagrams* which are just finite
order ideals in $\mathbb{N} \times \mathbb{N}$, typically designated by Greek letters. This is to say
that μ is a Young diagram if and only if for any (k, l) in μ and $(i, j) \leq (k, l)$
we must have (i, j) in μ. In other words, any cell lying to the southwest
of a cell of μ is also in μ. Clearly a Young diagram is characterized by the
decreasing integer sequence $(\mu_1, \mu_2, \ldots, \mu_k)$ with μ_i denoting the number
of cells in the ith row of μ. This will be further discussed in Section 1.6.

   The *arm* of a cell c of a Young diagram μ, is the set of cells of μ that
lie in the same row to the right of c (excluding c). Likewise, the *leg*[6] of a
cell is the set of cells that are above this cell and in the same column. The
arm length $a(c) = a_\mu(c)$ (resp. the *leg length* $\ell(c) = \ell_\mu(c)$) of c in μ is the
number of cells in the arm (resp. leg) of c in μ. Clearly for $c = (i, j)$ we
have

$$a_\mu(c) = \mu_{j+1} - i - 1 \quad \text{and} \quad \ell_\mu(c) = \mu'_{i+1} - j - 1.$$

The *hook* associated with a cell c of μ is the set of cells of μ lying in either
the arm or the leg of c as well as the cell c itself. The corresponding *hook
length* is $h(c) = a(c) + \ell(c) + 1$, so that we have

$$h(i, j) = \mu_{j+1} + \mu'_{i+1} - i - j - 1. \tag{1.4}$$

These notions are illustrated in Figure 1.10 for cell $(2, 2)$, with the arm in
yellow, and the leg in green.

   Let us underline again that we are following the right-side up "French"
convention, rather than the upside-down "English" convention, for drawing
diagrams, thus, $(0, 0)$ sits at the bottom left.

---

[6]The terminology comes from the English/American convention that insists on drawing diagrams upside down.

Figure 1.10. Arm, leg and hook of cell $(2, 2)$ of a Young diagram.

1.6 Partitions

Young diagrams usually appear in association with "partitions" of an integer. For simplicity's sake, we also denote by μ the *partition* associated with a Young diagram μ. This is just the decreasing ordered sequence $\mu = (\mu_1, \mu_2, \ldots, \mu_k)$ of row lengths of the diagram. Each μ_i is said to be a *part* of μ, and the sum $n = \mu_1 + \mu_2 + \cdots + \mu_k$ of these parts is evidently the number of cells of the diagram μ. We write $\mu \vdash n$ to indicate that μ is a partition of n. The *length* $\ell(\mu)$ of μ is just the number of (nonzero) parts of μ.

Partitions are often presented as *words* $\mu = \mu_1 \mu_2 \cdots \mu_k$ whose letters (parts) are integers > 0. We consider the empty partition as an exception to this no zero part rule, denoting it by 0. With this convention, we can present the *partition set* $\mathcal{P}(n)$ as

$$\mathcal{P}(0) = \{0\}$$
$$\mathcal{P}(1) = \{1\}$$
$$\mathcal{P}(2) = \{2, 11\}$$
$$\mathcal{P}(3) = \{3, 21, 111\}$$
$$\mathcal{P}(4) = \{4, 31, 22, 211, 1111\}$$
$$\mathcal{P}(5) = \{5, 41, 32, 311, 221, 2111, 11111\}$$
$$\mathcal{P}(6) = \{6, 51, 42, 411, 33, 321, 3111, 222, 2211, 21111, 111111\}$$
$$\vdots$$

Another common description of partitions in word format consists of writing[7] $\mu = \mathbf{1}^{m_1} \mathbf{2}^{m_2} \cdots \mathbf{j}^{m_j}$, with m_i equal to the number of parts of size i in μ. This is handy when describing the *cycle structure* or *shape* $\lambda(\sigma) = \mu$ of a permutation σ, with m_i being equal to the number of length i-cycles appearing in the cycle decomposition of σ. The classical result here is that two permutations σ and τ are conjugate if and only if they have the same shape, i.e., $\lambda(\sigma) = \lambda(\tau)$. Moreover, the number of shape μ permutations

---

[7] Here we use boldface to emphasize the difference between the word $\mathbf{k}^m = k\, k \cdots k$, consisting of m copies of the "letter" k, and the usual integer exponentiation k^m.

Figure 1.11. Corners of 4421.

in $\mathfrak{S}_n$ is $n!/z_\mu$, with $z_\mu := 1^{m_1} m_1! \, 2^{m_2} m_2! \cdots j^{m_j} m_j!$. We often encounter this number z_μ in later chapters.

For a partition μ of n, we say that the cell $c = (\mu_i - 1, i - 1)$ is a *corner* of μ if $\mu_i > \mu_{i+1}$. Corners are exactly the cells that can be removed so that the resulting diagram is still a Young diagram. For example, the corners of the partition $\mu = 4421$ are the three blue cells in Figure 1.11.

Operations on Partitions

The *sum* $\lambda + \mu$, of two partitions λ and μ, is the partition whose parts are equal to $\lambda_i + \mu_i$, with i going up to the maximum between the number of parts of λ and number of parts of μ. If need be, we consider parts $\lambda_i = 0$ and likewise for μ. This will be a running convention whenever it makes sense to do so. Using conjugation, we further define the *union* $\lambda \cup \mu := (\lambda' + \mu')'$. Equivalently, $\lambda \cup \mu$ is obtained by taking all the parts of λ jointly with those of μ (adding up multiplicities) and rearranging all these parts in descending size order. As illustrated in Figure 1.12 with $\lambda = 43111$ and $\mu = 211$, we obtain $\lambda + \mu = 64211$, whereas $\lambda \cup \mu = 43211111$ (not shown here).

Figure 1.12. $\lambda + \mu = 64211$.

Enumerating Partitions

Although there is no known simple formula for the number $p(n)$ of partitions of a given integer n, one has the following nice classical generating function:

$$\sum_{n \geq 0} p(n) x^n = \prod_{k \geq 1} \frac{1}{1 - x^k}. \tag{1.5}$$

To obtain it, we first expand each term of the right-hand side of (1.5) as a geometrical series

$$\frac{1}{1-x^i} = \sum_{m_i \geq 0} x^{im_i}. \tag{1.6}$$

Thus in the expansion of the product of (1.5) the term x^n appears exactly as often as there are solutions $(m_1, m_2, \dots)$ to the equation $n = \sum_i i m_i$. These solutions clearly correspond bijectively to a unique partition. The first terms of the resulting series are

$$\sum_{n \geq 0} p(n)x^n = 1 + x + 2x^2 + 3x^3 + 5x^4 + 7x^5 + 11x^6 + 15x^7 + 22x^8$$

$$+ 30x^9 + 42x^{10} + 56x^{11} + 77x^{12} + 101x^{13} + \cdots.$$

It is easy to adapt this argument to show that the generating function for the number of partitions with all parts distinct is

$$\sum_{n \geq 0} p_{\neq}(n)x^n = \prod_{k \geq 1}(1 + x^k). \tag{1.7}$$

We can also extend both of these formulas to account for partition enumeration with parts restricted to lie in some given subset of $\mathbb{N}$, or to have at most some given multiplicity. In the next section we recall a nice recursive formula due to Euler for the computation of $p(n)$.

Euler's Pentagonal Theorem

In order to state Euler's formula we need the notion of *pentagonal numbers*. These are the numbers $\omega(k) := k(3k-1)/2$ with k varying in $\mathbb{Z}$.

k	0	1	-1	2	-2	3	-3	4	-4	5	-5	$\cdots$
$\omega(k)$	0	1	2	5	7	12	15	22	26	35	40	$\cdots$

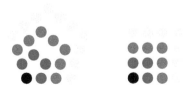

Figure 1.13. Pentagonal numbers and squares.

Considering only $k > 1$, we have the following rationale for this "pentagonal" adjective. A direct enumeration of the points in a side k "discrete regular pentagon" (see Figure 1.13) is easily seen to be given by

$$\sum_{i=0}^{k-1} (3i + 1) = \omega(k),$$

just as the number of points in a side k "discrete square" is given by

$$\sum_{i=0}^{k-1} (2i + 1) = k^2.$$

Euler showed that

Theorem 1.1 (Euler). *One has the equality*

$$\prod_{m \geq 1} (1 - x^m) = 1 + \sum_{k \geq 1} (-1)^k (x^{\omega(k)} + x^{\omega(-k)}), \tag{1.8}$$

from which is derived the recurrence

$$p(n) = \sum_{k \geq 1} (-1)^{k+1} \Big(p\big(n - \omega(k)\big) + p\big(n - \omega(-k)\big) \Big).$$

Proof: The following beautiful combinatorial proof of (1.8) is due to F. Franklin [Franklin 1881]. We start with a slight modification of formula (1.7):

$$\sum_{n,k \geq 0} p_{\neq}(n, k) q^k x^n = \prod_{m \geq 1} (1 + q x^m), \tag{1.9}$$

with $p_{\neq}(n, k)$ standing for the number of partitions of n into k distinct parts. Now let $p_{\neq}^+(n)$ (resp. $p_{\neq}^-(n)$) be the number of even length (resp. odd length) partitions of n into distinct parts. We write $\mathcal{P}_{\neq}^+(n)$ and $\mathcal{P}_{\neq}^-(n)$ for the corresponding sets of partitions. Clearly $p_{\neq}(n) = p_{\neq}^+(n) + p_{\neq}^-(n)$ and we get

$$\prod_{m \geq 1} \big(1 + (-1)x^m\big) = \sum_{n \geq 0} \sum_{k=0}^{n} (-1)^k p_{\neq}(n, k) x^n$$

$$= \sum_{n \geq 0} \big(p_{\neq}^+(n) - p_{\neq}^-(n)\big) x^n. \tag{1.10}$$

Figure 1.14. Bijection.

Thus (1.8) boils down to $p_{\neq}^+(n)$ being equal to $p_{\neq}^-(n)$ in most situations, the exceptions occurring when n is a pentagonal number, i.e., $n = \omega(\pm k)$. In these cases, we will see that $p_{\neq}^+(n) - p_{\neq}^-(n) = (-1)^k$. Indeed, we can exhibit a bijection between $P_{\neq}^+(n)$ and $P_{\neq}^-(n)$, whenever n is not of the form $\omega(\pm k)$. For the exceptional cases, the bijection is defined up to one special partition.

Drawing cells with "points" rather than squares, we identify two features of the Young diagram associated with a partition μ (see Figure 1.14). The first of these special features is just the (smallest) topmost part μ_ℓ of $\mu = (\mu_1, \ldots, \mu_\ell)$. The second special feature is the *prow* of μ obtained as follows. Let k be the largest integer such that $\mu_j = \mu_1 - j + 1$ for all $j \leq k$, then the prow of μ is the set of rightmost points in each of the first k rows of μ. Forgetting exceptional cases for the moment, we apply the following recipe to construct a new partition $\varphi(\mu)$ from μ by moving points between the topmost part and the prow of μ. Let $m := \mu_\ell$.

1. If $m \leq k$, then the topmost part of μ is removed and one point is added to each of the first m rows. This is the left-to-right correspondence in Figure 1.14.

2. Conversely, if $m > k$, a new topmost part of size k is added and one point is removed from each of the first k parts. This is the right-to-left correspondence in Figure 1.14.

In most cases the result of this construction is a partition with all parts distinct, exceptions occurring when the prow shares a point with the topmost part and $m - k \leq 1$. As illustrated in Figure 1.15, this forces n to be of the form $n = \omega(\pm k)$; moreover, the unique partition for which the bijection cannot be defined[8] is either $(2k - 1, \ldots, k + 1, k)$ or $(2k, \ldots, k + 2, k + 1)$. Both are clearly of parity $(-1)^k$. This completes the proof of (1.8). $\square$

---

[8] Because the result is not a partition, or is not a partition having distinct parts.

1st case   2nd case

Figure 1.15. Exceptions.

1.7   The Young Lattice

The next interesting notion arises when we order partitions by *inclusion*, considering partitions as subsets of $\mathbb{N} \times \mathbb{N}$. The resulting partial order is a *lattice* with union as *meet* and intersection as *join*. It is most often called the *Young lattice* and we denote it by $\mathcal{Y}$. For a partition μ to cover a partition ν in $\mathcal{Y}$ it must differ from ν by exactly one cell which has to be a corner of μ. We write $\nu \to \mu$ when this is the case. The bottom part of Young's lattice is depicted in Figure 1.16. For $\alpha \subseteq \beta$, the *interval* $[\alpha, \beta]$ is the set $\{\mu \mid \alpha \subseteq \mu \subseteq \beta\}$. An interesting special case is the interval $[0, \mathbf{n}^k]$ of partitions contained in the *rectangular* partition $\mathbf{n}^k$. The number of such partitions is given by the binomial coefficient $\binom{n+k}{k}$. This enumeration can be refined to give a weighted enumeration formula for the partitions lying in this interval giving the classical q-analog of the binomial coefficient:

$$\sum_{\mu \subseteq n^k} q^{|\mu|} = \begin{bmatrix} n + k \\ k \end{bmatrix}_q, \tag{1.11}$$

This notion will be studied in more detail in the next subsection.

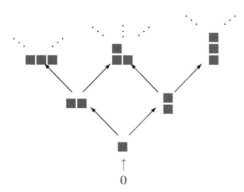

Figure 1.16. Young's lattice.

q-Binomial Coefficients

The notion of q-*analog* has a long history, going back to at least the 19th century with Heine's study of basic hypergeometric series. The general idea is to derive identities involving a parameter q that specialize to well known expressions when q tends to 1. In combinatorial settings, both sides of the identity are positive integer expressions in q, which specialize to combinatorially interesting integers when we set $q = 1$. One starts with the observation that

$$\lim_{q \to 1} \frac{1 - q^n}{1 - q} = \lim_{q \to 1} 1 + q + \cdots + q^{n-1} = n,$$

and writes $[n]_q$ for $1 + q + \cdots + q^{n-1}$. We have already encountered the q-analog of $n!$ (see (1.1), p. 7):

$$[n]_q! = [1]_q[2]_q \cdots [n]_q.$$

This is also written in the form $(q; q)_m$ using the notion of the q-*shifted factorial*

$$(a; q)_m := \begin{cases} (1 - a)(1 - aq) \cdots (1 - aq^{m-1}) & \text{if } m > 0, \\ 1 & \text{if } m = 0. \end{cases}$$

The next step is to mimic the usual expression for binomial coefficients in terms of factorials and define the q-*binomial coefficients*

$$\begin{bmatrix} m \\ k \end{bmatrix}_q := \frac{[m]_q!}{[k]_q![m - k]_q!}. \tag{1.12}$$

By giving a proof of (1.11) we will have shown that the right-hand side of (1.12) simplifies to a positive integer polynomial. This is confirmed by computing small values giving the beginning of the q-Pascal triangle:

$$1$$
$$1 \qquad 1$$
$$1 \qquad q + 1 \qquad 1$$
$$1 \qquad q^2 + q + 1 \qquad q^2 + q + 1 \qquad 1$$
$$1 \qquad q^3 + q^2 + q + 1 \qquad q^4 + 2q^2 + q^3 + q + 1 \qquad q^3 + q^2 + q + 1 \qquad 1$$

which can be prolonged using the linear recurrence

$$\begin{bmatrix} m \\ k \end{bmatrix}_q = q^k \begin{bmatrix} m - 1 \\ k \end{bmatrix}_q + \begin{bmatrix} m - 1 \\ k - 1 \end{bmatrix}_q, \tag{1.13}$$

with initial conditions $\begin{bmatrix} m \\ k \end{bmatrix}_q = 1$, if $k = 0$ or $k = m$. We will show that this recurrence is satisfied by both sides of (1.11) thus proving the equality.

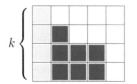
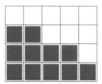

Figure 1.17. The two possibilities for the right-hand side of (1.14).

Formula (1.12) is easily checked by a direct computation. On the other hand, the weighted enumeration of partitions μ contained in a rectangle $\mathbf{n}^k$ corresponds to setting m equal to $n + k$ in (1.13), and we split the calculation depending on whether μ has exactly k parts or less. In the first case we remove the first column of μ to get a partition contained in the rectangle $(\mathbf{n} - \mathbf{1})^k$. The resulting partition has a weight that differs from that of μ by a factor of q^k. In the second case μ is already contained in the rectangle $\mathbf{n}^{k-1}$. Summing, we get

$$\sum_{\mu \subseteq \mathbf{n}^k} q^{|\mu|} = q^k \sum_{\nu \subseteq (\mathbf{n}-\mathbf{1})^k} q^{|\nu|} + \sum_{\mu \subseteq \mathbf{n}^{k-1}} q^{|\mu|}, \tag{1.14}$$

as illustrated in Figure 1.17. Initial conditions are verified directly. The following beautiful q-analog of Newton's binomial formula may sharpen the readers interest in q-binomial coefficients:

$$\prod_{k=0}^{n-1}(1 + q^k z) = \sum_{k=0}^{n} q^{k(k-1)/2} \begin{bmatrix} n \\ k \end{bmatrix}_q z^k. \tag{1.15}$$

Among other interesting binomial coefficient related q-analogs, we should certainly mention the striking fact that

$$\mathbf{C}_n(q) := \frac{1}{[n+1]_q} \begin{bmatrix} 2n \\ n \end{bmatrix}_q \tag{1.16}$$

is a positive integer coefficient polynomial. Neither the fact that $\mathbf{C}_n(q)$ is actually a polynomial, nor that it has positive integer coefficients, is evident. We will not give a proof of this well-known fact but we will give an explicit interpretation (in Section 10.5) for $\mathbf{C}_n(q)$ that will make it clear that it lies in $\mathbb{N}[q]$. The first values of $\mathbf{C}_n(q)$ are

$\mathbf{C}_1(q) = 1$,
$\mathbf{C}_2(q) = q^2 + 1$,
$\mathbf{C}_3(q) = q^6 + q^4 + q^3 + q^2 + 1$,
$\mathbf{C}_4(q) = q^{12} + q^{10} + q^9 + 2q^8 + q^7 + 2q^6 + q^5 + 2q^4 + q^3 + q^2 + 1$.

Setting $q = 1$ in (1.16) gives the classical *Catalan numbers*

$$C_n := \frac{1}{n+1}\binom{2n}{n}r,$$

so that formula (1.16) is said to give rise to a q-analog of the Catalan numbers. As we will see next, there is another natural and interesting q-analog of the Catalan numbers. This is a typical situation in the world of of q-analogs. Indeed, the construction of some q-analogs depend strongly on the choice of approach to the numbers we intend to q-analogize: e.g., explicit formula, recurrence, identity, and so on.

Exercise. For $k < \ell$ and $n < m$, find an expression for the q-enumeration of partitions μ such that $\mathbf{n}^k \subseteq \mu \subseteq \mathbf{m}^\ell$.

Partitions Contained in a Staircase

Another interesting interval in $\mathcal{Y}$ is the set of partitions contained in a *staircase partition* $\delta_k := (k-1, k-2, \ldots, 3, 2, 1)$ (see Figure 1.18). Again, the number of such partitions is the Catalan number for which the relevant property here is to satisfy the well-known recurrence

$$C_{k+1} = \sum_{j=0}^{k} C_j C_{k-j} \tag{1.17}$$

with initial condition $C_0 = 1$. Indeed, our forthcoming argument will actually prove that for $C_k(q) := \sum_{\mu \subseteq \delta_k} q^{|\mu|}$, we have

$$C_{k+1}(q) = \sum_{j=0}^{k} q^{j(k-j)} C_j(q) C_{k-j}(q). \tag{1.18}$$

To show it we use *Dyck paths*. These are sequences of *steps* $(a_i, b_i) \to (a_{i+1}, b_{i+1})$ in $\mathbb{N} \times \mathbb{N}$ with

$$(a_{i+1}, b_{i+1}) = \begin{cases} (a_i, b_i) + (1, 0) & \text{or} \\ (a_i, b_i) - (0, 1) \end{cases}$$

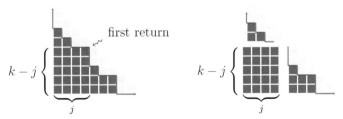

Figure 1.18. From partitions to Dyck paths.

going from $(0, k)$ to $(k, 0)$, and such that all (a_i, b_i) remain inside the staircase shape, i.e., $a_i + b_i \le k$. Figure 1.18 illustrates how the "boundary" of a partition, contained in the staircase δ_k, corresponds to a Dyck path. It also shows how to decompose both the partition and its associated Dyck path taking into account the "first return to the diagonal". In the path outlook, this is the smallest $i > 0$ for which we have $(a_i, b_i) = (j, k - j)$, with $1 \le j \le k$. The tail of the path, starting at this point $(j, k - j)$ corresponds to Dyck path, up to a left shift by j. We also get a (general) Dyck path out of the portion going from $(0, k)$ to $(j, k - j)$ after removing the first and last steps. This procedure is clearly bijective. It translates into the q-identity (1.18) if we multiply each term by $q^{j(k-j)}$ to conserve the weight.

The first values of this alternate q-analog are

$$\mathcal{C}_1(q) = 1,$$
$$\mathcal{C}_2(q) = q + 1,$$
$$\mathcal{C}_3(q) = q^3 + 2\,q^2 + q + 1,$$
$$\mathcal{C}_4(q) = q^6 + 3\,q^5 + 3\,q^4 + 3\,q^3 + 2\,q^2 + q + 1,$$
$$\mathcal{C}_5(q) = q^{10} + 4\,q^9 + 6\,q^8 + 7\,q^7 + 7\,q^6 + 5\,q^5$$
$$+ 5\,q^4 + 3\,q^3 + 2\,q^2 + q + 1.$$

Comparing with the values of $\mathbf{C}_n(q)$ given previously, we see that we have two different q-analogs of the Catalan numbers.

Skew Partitions

For any partitions μ and λ such that $\lambda \subseteq \mu$, we define the *skew* partition μ/λ to be the diagram obtained as the set difference of μ and λ. This is illustrated in Figure 1.19 with the skew partition $755321/5421$. If (as in Figure 1.20) no two cells of μ/λ lie in the same column we say that μ/λ is a *horizontal strip*. In a similar manner, if no two cells of μ/λ lie in the same row we say that it is a *vertical strip*. If, as in Figure 1.19, the skew partition is connected and contains no 2×2 squares, we say that we have a *ribbon*. One can easily check that (up to translation) there are exactly as many n cell ribbons as there are "compositions" of n (see Section 1.8).

$$755321/5421 =$$

Figure 1.19. A skew partition.

Figure 1.20. A horizontal strip.

Orders on Partitions

Two orders play an important role for partitions of n. First and foremost, we have the (partial) *dominance* order $\lambda \preceq \mu$. We say that μ *dominates* λ if and only if, for all k, $\lambda_1 + \lambda_2 + \cdots + \lambda_k \leq \mu_1 + \mu_2 + \cdots + \mu_k$. If needed, parts $\mu_i = 0$ or $\lambda_i = 0$ can be added so that inequalities make sense. Figure 1.21 gives the dominance order on partitions of $n = 6$, with an arrow $\mu \to \lambda$ indicating that μ is covered by λ in the dominance order. This example underlines the fact that the dominance order is not a total order. Another useful order on partitions is the *lexicographic* order. We say $\lambda < \mu$ if the first nonzero difference $\mu_i - \lambda_i$ is positive. In increasing lexicographic order, the partitions of 6 are listed in Figure 1.22.

Exercise. Show that the lexicographic order on a partition is a linear extension of the dominance order, i.e., $\lambda \preceq \mu$ implies that $\lambda \leq \mu$.

Exercise. Show that the dominance order is "symmetric" with respect to conjugation. This is to say that $\lambda \prec \mu$ if and only if $\mu' \prec \lambda'$.

Exercise. Show that the function $n(\mu)$ (see (1.2)) is decreasing for the dominance order. This is to say that $n(\mu) > n(\lambda)$ if $\mu \prec \lambda$.

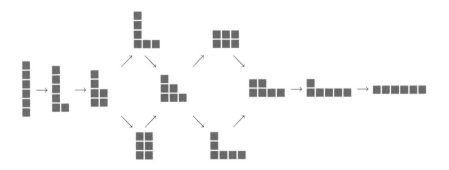

Figure 1.21. Dominance order for $n = 6$.

$$111111 \; \leq \; 21111 \; \leq \; 2211 \; \leq \; 222 \; \leq \; 3111 \; \leq$$
$$\leq \; 321 \; \leq \; 33 \; \leq \; 42 \; \leq 411 \; \leq \; 51 \; \leq \; 6.$$

Figure 1.22. Lexicographic order for $n = 6$.

1.8   Compositions

A *composition* $\mathbf{c} = (c_1, c_2, \ldots, c_k)$ of an integer n is an ordered sequence of *parts* $c_i > 0$ (in $\mathbb{N}$) that sum up to n. We write $\mathbf{c} \models n$ when n is a composition of n. The *length* $\ell(\mathbf{c})$ of a composition $\mathbf{c}$ is the number of its parts. We can readily check that there are 2^{n-1} compositions of n, by constructing a bijective correspondence between composition of n and subsets of $\{1, 2, \ldots, n-1\}$. To do this, we associate with the composition $\mathbf{c} = (c_1, c_2, \ldots, c_k)$ the set $S_{\mathbf{c}} := \{s_1, s_2, \ldots, s_{k-1}\}$, with $s_i = c_1 + \cdots + c_i$. For given n, this process is entirely reversible since $c_i = s_i - s_{i-1}$, with the understanding that $s_0 = 0$ and $s_k = n$. The composition of n that is associated in this manner with a subset T of $\{1, \ldots, n-1\}$ is denoted by $\mathrm{co}(T)$. In particular, this implies that there are $\binom{n-1}{k-1}$ length k compositions of n. One interesting use of this correspondence is to associate with a permutation σ the *descent composition* $\mathrm{co}(\sigma)$ encoding the descent set $\mathrm{Des}(\sigma)$.

The *refinement order* between compositions of n corresponds to reverse inclusion of the associated sets, so that $\mathbf{a} \succeq \mathbf{b}$ if and only if $S_{\mathbf{a}} \subseteq S_{\mathbf{b}}$. The composition $\mathbf{b}$ is obtained from $\mathbf{a}$ by splitting some of its parts. For example, $(7, 3, 8)$ is thus obtained from $(2, 5, 3, 3, 2, 3)$, since $7 = 2+5$, $3 = 3$, and $8 = 3 + 2 + 3$. The associated subsets are $\{7, 10\}$ and $\{2, 7, 10, 13, 15\}$ and we indeed have the required containment.

We denote by $\widehat{\mathbf{c}}$ the partition obtained by sorting the parts of $\mathbf{c}$ in decreasing order. This is the *form* of $\mathbf{c}$. It is easy to check that there are $k!/(m_1! \, m_2! \cdots m_j!)$ length $k = m_1 + m_2 + \cdots + m_j$ compositions of form equal to the partition $1^{m_1} 2^{m_2} \cdots j^{m_j}$. The generating function for length k compositions is just

$$\left(\frac{q}{1-q} \right)^k.$$

For our exposition, it may sometimes be useful to add some 0 parts to length k ($\leq n$) compositions, to turn them into n-cells in $\mathbb{N}^n$. This allows us to make sense of an identity such as $(1, 3, 1, 2) + (1, 1, 2, 2, 1, 2) = (2, 4, 3, 4, 1, 2)$. In more compact form this may also be written as $1312 + 112212 = 243412$, if it is clear that we are dealing with compositions.

1.9 Words

Let us first choose some *alphabet* (any finite set) $\mathbf{A}$ whose elements are called *letters*. A *word* u on $\mathbf{A}$ is just a (finite) sequence $u = a_1 a_2 \cdots a_k$ of letters a_i in the alphabet $\mathbf{A}$. We include here the *empty* word ε. The set $\mathbf{A}^*$, of all words on $\mathbf{A}$, is a *monoid* (or semigroup) for the *concatenation* operation defined as $u \cdot v := a_1 a_2 \cdots a_k b_1 b_2 \cdots b_m$ when $u = a_1 a_2 \cdots a_k$ and $v = b_1 b_2 \cdots b_m$. This is clearly an associative operation for which the empty word ε acts as identity. Compositions are often written as words, and their concatenation is a natural operation such that that $\mathbf{a} \cdot \mathbf{b}$ is a composition of $n + k$ if $\mathbf{a}$ is a composition of n and $\mathbf{b}$ is a composition of k.

We say that $\ell(u) := k$ is the *length* of $u = a_1 a_2 \cdots a_k$, setting 0 to be the length of the empty word. We recursively set $u^n := u(u^{n-1})$ for $n > 0$, with $u^0 := \varepsilon$. Thus u^n is a word of length nk. If $\mathbf{A}$ has cardinality n, then the set $\mathbf{A}^k$ of length k words is of cardinality n^k, and we have $\mathbf{A}^* = \sum_{k \geq 0} \mathbf{A}^k$. The *word algebra* $\mathbb{Q}\mathbf{A}^*$ is the free vector space on $\mathbf{A}^*$ over the field[9] $\mathbb{Q}$, so that its elements are finite linear combinations of words, on which concatenation is extended bilinearly. The algebra $\mathbb{Q}\mathbf{A}^*$ is naturally graded by word length, so that we have the direct sum decomposition $\mathbb{Q}\mathbf{A}^* = \bigoplus_{k \geq o} \mathbb{Q}\mathbf{A}^k$. The *shuffle* operation $\sqcup\!\sqcup$ is a bilinear transformation on $\mathbb{Q}\mathbf{A}^*$, recursively defined between words as

$$u \sqcup\!\sqcup v = \begin{cases} u & \text{if } v = \varepsilon, \\ v & \text{if } u = \varepsilon, \\ a(u' \sqcup\!\sqcup v) + b(u \sqcup\!\sqcup v') & \text{if } u = au' \text{ and } v = bv', \, a, b \in \mathbf{A}. \end{cases}$$

For example, the shuffle of $\mathbf{12}$ and $\mathbf{34}$ is the sum of the 6 words

$$\mathbf{12} \sqcup\!\sqcup \mathbf{34} = \mathbf{1234} + \mathbf{1324} + \mathbf{1342} + \mathbf{3124} + \mathbf{3142} + \mathbf{3412}.$$

Each of these 4-letter words corresponds to a possible choice for the insertion of $\mathbf{12}$ as a subword. This goes to illustrate that the number of terms in the shuffle of two words, of respective lengths k and m, is the binomial coefficient $\binom{k+m}{m}$. Observe that it may happen that words appear with multiplicity in a shuffle product. For instance,

$$\mathbf{12} \sqcup\!\sqcup \mathbf{12} = 2 \cdot \mathbf{1212} + 4 \cdot \mathbf{1122}.$$

In a word of the form $w = uvt$, we say that u, v, and t are respectively a *prefix*, a *factor*, and a *suffix* of the word w. Notice that we have not excluded the possibility that one or more of these is the empty word. If the alphabet is ordered, just as for permutations we can define for words the notions of *descent*, *inversion*, *major index*, etc.

---

[9]The underlying field is most often $\mathbb{Q}$, but we leave open the possibility of considering other characteristic 0 fields.

Standardization

Let $w = a_1 a_2 \cdots a_n$ be an n-letter word on an ordered alphabet, say $\mathbb{N}$. The *standardization* $\mathrm{st}(w)$, of w, is the unique permutation $\sigma \in \mathfrak{S}_n$ such that $\sigma_i < \sigma_j$ whenever we either have $a_i < a_j$, or $a_i = a_j$ and $i < j$. Thus for $w = 2241321$ we get $\mathrm{st}(w) = 3471652$. A somewhat "reverse" process for standardization is that of minimization. The *minimization* $\mathrm{mn}(u)$ of a word u is the smallest word (in lexicographic order) whose standardization coincides with that of u. In the case of permutations, it can alternatively be described as

$$\mathrm{mn}(\sigma) := \mathbf{0}^n + \sum_{i \in \mathrm{Des}(\sigma^{-1})} (\mathbf{e}_{\sigma^{-1}(i+1)} + \cdots + \mathbf{e}_{\sigma^{-1}(n)}), \qquad (1.19)$$

with the $\mathbf{e}_k$ denoting the usual standard basis vectors of $\mathbb{N}^n$. To illustrate, consider the permutation $\sigma = 3471652$, whose inverse $\sigma^{-1} = 4712653$ has descent set $\{2, 5, 6\}$. We get

$$\mathrm{mn}(3471652) = 0000000 + (\mathbf{e}_1 + \mathbf{e}_2 + \mathbf{e}_6 + \mathbf{e}_5 + \mathbf{e}_3) + (\mathbf{e}_5 + \mathbf{e}_3) + \mathbf{e}_3$$
$$= 1130210,$$

and we check that $\mathrm{st}(1130210) = 3471652$. Observe that for $u = 4491753$, we also have $\mathrm{mn}(u) = 1130210$. A general procedure for writing down $\mathrm{mn}(u)$, consists of successively replacing the letters of u as follows. One reads the letters of u from the smallest to the largest (and from left to right among equal letters). Each letter is replaced by the current value of a "counter" whose value starts at 0 and goes up by one each time we move to the left in u to read the next letter. Thus as long as we encounter equal values or we go right, we replace the letters with the same current value of the counter.

Exercise. Show that the descent set of the word u is the same as the descent set of the word $\mathrm{mn}(u)$.

We will see in Section 2.6 that the sum of the letters of $\mathrm{mn}(u)$ plays a role in the notion of "cocharge" of tableaux. Just for this, we denote by $\mathrm{coch}(u)$ this sum of letters, and call it the *cocharge* of the word u.

Row Decomposition

Consider an alphabet $\mathbf{A}$ with a total order on the letters (for example, $\{1, 2, \ldots, n\}$ with its usual order). A nondecreasing word $w = a_1 a_2 \cdots a_m$ is a word in which the letters increase from left to right, i.e., $a_1 \leq a_2 \leq \cdots \leq a_m$. The *row decomposition* of a word u is the unique decomposition of u as a product of maximal nondecreasing words: $u = w^{(1)} w^{(2)} \cdots w^{(k)}$.

Each $w^{(i)}$ is said to be a *row* of u. Thus, the row $w^{(i)}$ is a nondecreasing word whose last letter is larger than the first letter of the next row $w^{(i+1)}$, $1 \leq i \leq k-1$. The motivation behind the terminology is that we are going to identify "tableaux" with some special words, and rows in tableaux will correspond to rows in words. Observe that in the row decomposition $u = w^{(1)} w^{(2)} \cdots w^{(k)}$ the word $w^{(k)}$ is the largest nondecreasing suffix of w. We illustrate all this with the row decomposition

$$86934462235 = \underbrace{8}_{w^{(1)}} \; \underbrace{69}_{w^{(2)}} \; \underbrace{3446}_{w^{(3)}} \; \underbrace{2235}_{w^{(4)}}.$$

Chapter 2

Some Tableau Combinatorics

One of the leitmotivs running through this book is the deduction of formulas from combinatorial manipulations on tableaux. Central to this is the Robinson–Schensted–Knuth correspondence (see [Knuth 70, Robinson 38, Schensted 61]) between pairs of words of the same length and pairs of semi-standard tableaux of the same shape. We can deduce many fundamental identities from this correspondence. For more details on the combinatorics of tableaux, we refer the reader to Chapter 6 of [Lothaire 02] or to [Blessenohl and Schocker 05]. Another excellent reference is [Fulton 97].

2.1 Tableaux

We now come to another central notion, that of tableaux. Among several other roles they play, a crucial one is the construction of irreducible representations of the symmetric group. The construction in question involves the calculation of certain polynomials associated with "fillings" of the cells of a diagram. Such fillings are called tableaux. More precisely, a *tableau* τ of *shape* $\mathbf{d}$ with values in a set $\mathbf{A}$ (usually some subset of $\mathbb{N}$) is a function $\tau : \mathbf{d} \longrightarrow \mathbf{A}$. We denote by $\lambda(\tau)$ the shape $\mathbf{d}$ of τ, and we think of $\tau(c)$ as being the *entry* or *value* that appears in cell c. We also say that c in $\mathbf{d}$ is a cell of τ. A tableau τ of shape $\mathbf{d}$ is *semi-standard*, if its entries are nondecreasing along rows (from left to right), and strictly increasing along columns (from bottom to top). This is to say that $i < k$ implies $\tau(i, j) \leq \tau(k, j)$ and $j < \ell$ implies $\tau(i, j) < \tau(i, \ell)$, whenever these statements make sense. A n-cell tableau τ is *standard* if it is a bijective semi-standard tableau with values in $\{1, 2, \ldots, n\}$. Thus, for τ to be standard we need $\tau(i, j) < \tau(k, \ell)$ whenever $(i, j) < (k, \ell)$ coordinatewise. Figure 2.2 gives an example of a standard tableau of shape 431. For a partition μ we can interpret standard tableaux of shape μ as maximal chains $\mu^{(0)} \to \mu^{(1)} \to \cdots \to \mu^{(n)}$, from

31

Figure 2.1. A tableau.   **Figure 2.2.** A standard tableau.

Figure 2.3. A maximal chain in Young's lattice.

$\mu^{(0)} = 0$ to $\mu^{(n)} = \mu$, in the Young lattice. The maximality property says that two consecutive partitions $\mu^{(i)} \subseteq \mu^{(i+1)}$ must differ by exactly one cell, which is a corner of $\mu^{(i+1)}$. For a tableau τ, the corresponding chain is obtained by setting $\mu^{(i)}$ to be the partition whose cells take values that are less than or equal to i. In other words, we consider the cell labeled i as the one that gets "added" at the ith step. For example, the standard tableau in Figure 2.2 corresponds (bijectively) to the maximal chain of Figure 2.3. This correspondence between standard tableaux and maximal chains extends easily to the context of standard tableaux of shape given by a skew partition μ/λ, the corresponding maximal chain going from λ to μ. There is also a correspondence between length k chains (not necessarily maximal) and surjective semi-standard tableaux with values in $\{1, \ldots, k\}$. In this last case, a horizontal strip is added at each step.

The *reading word* $\omega(\tau)$ of a tableau τ is the sequence of letters $\omega(\tau) = \tau(c_1)\tau(c_2) \cdots \tau(c_n)$, appearing in cells of τ in their reading order (see Section 1.6). Recall that this means that we "read" the entries of τ with the usual conventions of the English (or French) language. For example, we have the reading word $\omega(\tau) = 86934462235$ for the tableau in Figure 2.4, with emphasis on the row decomposition $w^{(1)}w^{(2)}w^{(3)}w^{(4)}$ of $\omega(\tau)$. This illustrates that the rows of $\omega(\tau)$ coincide with the rows of τ. The underlying

$$\omega \begin{pmatrix} 8 \\ 6 \ 9 \\ 3 \ 4 \ 4 \ 6 \\ 2 \ 2 \ 3 \ 5 \end{pmatrix} = \underbrace{8}_{w^{(1)}} \ \underbrace{69}_{w^{(2)}} \ \underbrace{3446}_{w^{(3)}} \ \underbrace{2235}_{w^{(4)}}.$$

Figure 2.4. Reading word of a semi-standard tableau.

correspondence gives a simple bijection between semi-standard tableaux and reading words. Among the 81 words of length 4 on the alphabet $\{1, 2, 3\}$ only the following 39 are reading words of semi-standard tableaux:

$1111, 1112, 1113, 1122, 1123, 1133, 1222, 1223, 1233, 1333, 2111, 2112, 2113,$

$2122, 2123, 2133, 2211, 2222, 2223, 2233, 2311, 2312, 2333, 3111, 3112, 3113,$

$3122, 3123, 3133, 3211, 3212, 3213, 3222, 3223, 3233, 3311, 3312, 3322, 3333.$

The Hook Length Formula

The number f^μ of standard tableaux of shape μ ($\mu \vdash n$) is given by the Frame–Robinson–Thrall hook length formula[1] (see [Frame et al. 54]).

$$f^\mu = \frac{n!}{\prod_{c \in \mu} h(c)}, \tag{2.1}$$

where $h(c)$ is the hook length of $c = (i - 1, j - 1)$ in μ. Recall that $h(c) = (\mu_j - i) + (\mu_i' - j) + 1$. A direct application of formula (2.1) says that there should be exactly 16 standard tableaux of shape 321. These are catalogued in Figure 2.5.

Exercise. Show that the sum of the hook lengths of a partition μ is given by the formula $\sum_{c \in \mu} h(c) = n(\mu) + n(\mu') + |\mu|$.

Kostka Numbers

The *content* $\gamma(\tau)$ of a tableau τ is the sequence $\gamma(\tau) = (m_1, m_2, m_3, \dots)$ of *multiplicities* of each entry i in the tableau τ. For example, the content of the semi-standard tableau

$$\begin{array}{ccccc} 4 & 4 & & & \\ 2 & 2 & 4 & 4 & \\ 1 & 1 & 1 & 1 & 2 \end{array}$$

is $\gamma(\tau) = (4, 2, 0, 4, 0, 0, \dots)$. Clearly the numbers m_i sum up to the number of cells of the underlying tableau. In general this sequence need not be a partition, but we are especially interested in situations for which this is the case. If λ and μ are both partitions of the same integer n, we define the *Kostka number* $K_{\lambda, \mu}$ to be the number of semi-standard tableaux of shape

---

[1]See Section 4.6 for a proof, as well as a formula for the number of semi-standard tableaux.

$$
\begin{array}{ccc}
3 & & \\
2 & 5 & \\
1 & 4 & 6
\end{array}
\qquad
\begin{array}{ccc}
3 & & \\
2 & 6 & \\
1 & 4 & 5
\end{array}
\qquad
\begin{array}{ccc}
4 & & \\
2 & 5 & \\
1 & 3 & 6
\end{array}
\qquad
\begin{array}{ccc}
4 & & \\
2 & 6 & \\
1 & 3 & 5
\end{array}
$$

$$
\begin{array}{ccc}
4 & & \\
3 & 5 & \\
1 & 2 & 6
\end{array}
\qquad
\begin{array}{ccc}
4 & & \\
3 & 6 & \\
1 & 2 & 5
\end{array}
\qquad
\begin{array}{ccc}
5 & & \\
2 & 4 & \\
1 & 3 & 6
\end{array}
\qquad
\begin{array}{ccc}
5 & & \\
2 & 6 & \\
1 & 3 & 4
\end{array}
$$

$$
\begin{array}{ccc}
5 & & \\
3 & 4 & \\
1 & 2 & 6
\end{array}
\qquad
\begin{array}{ccc}
5 & & \\
3 & 6 & \\
1 & 2 & 4
\end{array}
\qquad
\begin{array}{ccc}
5 & & \\
4 & 6 & \\
1 & 2 & 3
\end{array}
\qquad
\begin{array}{ccc}
6 & & \\
2 & 4 & \\
1 & 3 & 5
\end{array}
$$

$$
\begin{array}{ccc}
6 & & \\
2 & 5 & \\
1 & 3 & 4
\end{array}
\qquad
\begin{array}{ccc}
6 & & \\
3 & 4 & \\
1 & 2 & 5
\end{array}
\qquad
\begin{array}{ccc}
6 & & \\
3 & 5 & \\
1 & 2 & 4
\end{array}
\qquad
\begin{array}{ccc}
6 & & \\
4 & 5 & \\
1 & 2 & 3
\end{array}
$$

Figure 2.5. All the standard tableaux of shape 321.

λ and content μ. For instance, the four possible semi-standard tableaux having content 2211 and shape 321 are

$$
\begin{array}{ccc}
4 & & \\
2 & 3 & \\
1 & 1 & 2
\end{array}
,
\qquad
\begin{array}{ccc}
3 & & \\
2 & 4 & \\
1 & 1 & 2
\end{array}
,
\qquad
\begin{array}{ccc}
4 & & \\
2 & 2 & \\
1 & 1 & 3
\end{array}
,
\qquad
\begin{array}{ccc}
3 & & \\
2 & 2 & \\
1 & 1 & 4
\end{array}
.
$$

Observe that the content of a semi-standard tableau is $\mathbf{1}^n$ if and only if the tableau is standard. This implies that $K_{\lambda,1^n} = f_\lambda$. We also easily deduce that $K_{\lambda,\lambda} = 1$, since a semi-standard tableau of content equal to its shape must have all cells in the ith row filled with the same value i. A straightforward extension of this argument shows that we must also have $K_{\lambda,\mu} = 0$ whenever μ is not smaller than λ in dominance order. Indeed, if $\mu_1 \geq \lambda_1$ then a copy of 1 must appear in the tableau outside of the first row, hence the tableau cannot be semi-standard. Likewise, when $\mu_1 + \mu_2 \geq \lambda_1 + \lambda_2$, either 1 or 2 appears in a row above the second one, implying again that the tableau is not semi-standard, etc.

The above observations give a somewhat more natural characterization of the dominance order. Indeed, we have $\lambda \succeq \mu$ if and only if $K_{\lambda,\mu} \neq 0$. Moreover, we see that the matrix $(K_{\lambda,\mu})_{\lambda,\mu \vdash n}$ is upper triangular when partitions are sorted in any decreasing order which is a linear extension of the dominance order. This immediately implies that the matrix $(K_{\lambda,\mu})_{\lambda,\mu \vdash n}$ is invertible. For $n = 4$ and using reverse lexicographic order, the Kostka matrix is

$$
\begin{matrix}
4 & 31 & 22 & 211 & 1111
\end{matrix}
$$

$$
\begin{pmatrix}
1 & 1 & 1 & 1 & 1 \\
0 & 1 & 1 & 2 & 3 \\
0 & 0 & 1 & 1 & 2 \\
0 & 0 & 0 & 1 & 3 \\
0 & 0 & 0 & 0 & 1
\end{pmatrix} . \tag{2.2}
$$

2.2   Insertion in Words and Tableaux

We now define an operation of insertion of a letter in a word closely related to an insertion of an entry in a tableau. More precisely, for a totally ordered alphabet $\mathcal{A}$, the *insertion* of a letter x in a word u is the word $w = (u \leftarrow x)$ recursively constructed as follows. If u is the empty word, then we just set $(u \leftarrow x) := x$. Otherwise, let $u = v\theta$ with θ the maximal length nondecreasing suffix of u, and set

$$
(u \leftarrow x) := \begin{cases} ux & \text{if } \max(\theta) \leq x, \\ (v \leftarrow y)\theta' & \text{otherwise.} \end{cases}
$$

Here y is the leftmost letter of θ that is larger than x so that $\theta = \alpha y\beta$, with $\max(\alpha) \leq x$, and θ' is then defined to be $\alpha x\beta$. In other words, y is in the rightmost position where we can substitute x for y in θ, so that the resulting word θ' is nondecreasing. We say that x *bumps* y from θ. To extend the insertion operation $(u \leftarrow v)$ to words, we successively insert the letters in v going from left to right. Thus, we force the relation

$$
(u \leftarrow vw) = ((u \leftarrow v) \leftarrow w). \tag{2.3}
$$

For instance, $(213 \leftarrow 21132412) = 33222411112$.

Exercise. Let τ be a semi-standard tableau and x a letter. Using a recursion on the number of cells in τ, show that $\omega(\tau) \leftarrow x$ is the reading word of some semi-standard tableau.

It follows that for a semi-standard tableau τ, the result of any insertion $\omega(\tau) \leftarrow u$ is the reading word of some semi-standard tableau τ'. In formula, $\big(\omega(\tau) \leftarrow u\big) = \omega(\tau')$.

Exercise. Show that for all semi-standard tableau τ we have $\big(\varepsilon \leftarrow \omega(\tau)\big) = \omega(\tau)$ by, checking that $\omega(\tau) = v \leftarrow \theta$ if we have the decomposition $\omega(\tau) = v\theta$ with θ being the relevant maximal nondecreasing suffix.

This observation makes it legitimate to consider an *insertion process* for tableaux. We denote by $(\tau \leftarrow v)$ the unique tableau such that $\omega(\tau \leftarrow v) = \big(\omega(\tau) \leftarrow v\big)$ holds, implicitly identifying a tableau τ with its reading word $\omega(\tau)$. For example, we have

$$
\begin{pmatrix} \begin{array}{cc} 2 & \\ 1 & 3 \end{array} & \leftarrow 21132412 \end{pmatrix} = \begin{array}{cccc} 3 & 3 & & \\ 2 & 2 & 2 & 4 \\ 1 & 1 & 1 & 1 & 2 \end{array}.
$$

The tableau obtained by successive insertion of the letters of a word u, starting with the empty tableau, i.e., $(0 \leftarrow u)$, is called the *insertion tableau* of u.

Exercise. Consider the skew partition obtained as the difference μ/λ

with μ equal to the shape of $(\tau \leftarrow v)$, for some semi-standard tableau τ of shape λ:

$$
\begin{pmatrix} \begin{array}{cccc} 2 & 5 & & \\ 1 & 3 & 4 & 6 \end{array} & \leftarrow 11235 \end{pmatrix} = \begin{array}{cccccc} 5 & & & & & \\ 2 & 3 & 4 & 6 & & \\ 1 & 1 & 1 & 2 & 3 & 5 \end{array}.
$$

Show that there is a bijection between pairs (τ, v), with τ a semi-standard tableau of shape λ and v nondecreasing, and pairs (τ', λ), where τ' is a tableau of shape μ such that μ/λ is a horizontal strip. **Hint:** it may be useful to consider the notion of *insertion path*, which is the set of cells that are modified during an insertion. Then for two values $x_1 < x_2$ to be inserted, observe that the insertion path for x_1 lies entirely to the left of the insertion path for x_2.

2.3   Jeu de Taquin

Let τ be a semi-standard tableau of shape $\mathbf{d}$. A *slide* move of *jeu de taquin* consists of one of the following local moves. A cell (rather its value) is either moved one step left or down, if the target cell is "empty", in other words, the target cell does not belong to $\mathbf{d}$. The following sliding rules (with empty cells marked in green) preserve the property of being semi-standard:

(1) $\begin{array}{c} a \\ b \end{array} \quad \rightsquigarrow \quad a \ b$, allowed if $a \leq b$ or the target row is empty,

(2) $\begin{array}{c} b \\ a \end{array} \quad \rightsquigarrow \quad \begin{array}{c} b \\ a \end{array}$, allowed if $a < b$ or the target column is empty.

A *rectification* of a skew semi-standard tableau τ consists of a sequence of slides for which the end shape is a partition. Thus, the result of a rectification is a semi-standard tableau. It is a fact (see [Fulton 97]) that the result of a rectification does not depend on the choice of sliding steps, so that it makes sense to talk about the *rectified* tableau of τ. An example of this rectification process is given in Figure 2.6 where, at each step, the green cells correspond to a sequence of cells that have been selected to be slid southwest, according to a sequence of applications of the sliding rules (1) and (2).

2.4 The Robinson–Schensted–Knuth (RSK) Correspondence

Let us introduce the following notations for "lexicographic words" of "biletters". A *biletter* $\binom{b}{a}$ is just another name and notation for an element (a, b) of the alphabet $\mathcal{A} \times \mathcal{B}$. Biletters are ordered lexicographically:

$$\binom{b}{a} \preceq \binom{d}{c} \quad \text{iff} \quad \begin{cases} b < d & \text{or} \\ b = d & \text{and} \quad a \leq c. \end{cases}$$

For $\mathbf{a} = (a_1, a_2, \ldots, a_k)$ and $\mathbf{b} = (b_1, b_2, \ldots, b_k)$ such that

$$\binom{b_1}{a_1} \preceq \binom{b_2}{a_2} \preceq \cdots \preceq \binom{b_k}{a_k},$$

we say that

$$\binom{\mathbf{b}}{\mathbf{a}} = \binom{b_1}{a_1}\binom{b_2}{a_2}\cdots\binom{b_k}{a_k} = \begin{pmatrix} b_1 & b_2 & \cdots & b_k \\ a_1 & a_2 & \cdots & a_k \end{pmatrix} \tag{2.4}$$

forms a *lexicographic word*. Notice that there are three possible notations here for a lexicographic word.

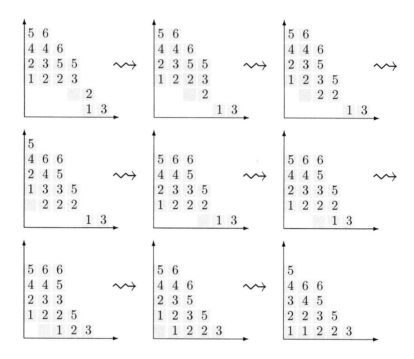

Figure 2.6. Rectification.

The RSK correspondence $u \longleftrightarrow (P, Q)$ associates bijectively a pair $(P(u), Q(u)) = (P, Q)$ of same shape semi-standard tableaux to each *lexicographic word*

$$u = \begin{pmatrix} b_1 & b_2 & \cdots & b_k \\ a_1 & a_2 & \cdots & a_k \end{pmatrix}. \tag{2.5}$$

The correspondence is recursively defined as follows.

(1) First, we set $\varepsilon \longleftrightarrow (0, 0)$.

(2) For a nonempty word $u = v \binom{b}{a}$, we recursively set $v \longleftrightarrow (P', Q')$ and construct

$$P := (P' \leftarrow a)$$

by tableau insertion of a into P'. The tableau Q is then obtained by adding b to Q' in the position given by the unique cell by which the shape of P differs from that of P'.

For instance, we have

$$
\begin{pmatrix}
3 & 5 & 5 & & & & & 3 & 4 & 4 & & & \\
2 & 3 & 3 & & & & & 2 & 2 & 2 & & & \\
1 & 1 & 1 & 2 & 3 & 3 & , & 1 & 1 & 1 & 1 & 2 & 2
\end{pmatrix}
\leftarrow \begin{pmatrix} 6 \\ 1 \end{pmatrix} =
$$

$$
\begin{pmatrix}
5 & & & & & & 6 & & & & & \\
3 & 3 & 5 & & & & 3 & 4 & 4 & & & \\
2 & 2 & 3 & & & & 2 & 2 & 2 & & & \\
1 & 1 & 1 & 1 & 3 & 3 , & 1 & 1 & 1 & 1 & 2 & 2
\end{pmatrix}.
$$

The RSK correspondence enjoys particular properties when the input word has some special structure. For example, starting with an "ordinary" word

$$ u = a_1 a_2 \cdots a_n $$

we can consider the canonical lexicographic biletter word

$$
\begin{pmatrix}
1 & 2 & \cdots & n \\
a_1 & a_2 & \cdots & a_n
\end{pmatrix}
$$

to which the RSK correspondence is applied to get the pair of tableaux (P, Q). Since the top line of the input word is $(1, 2, \ldots, n)$, it follows that the tableau Q is actually standard. In the even more special case when $a_1 a_2 \cdots a_n$ is a permutation of $\{1, 2, \ldots, n\}$, the tableau P is also standard. This establishes a bijection between permutations in $\mathfrak{S}_N$ and pairs of standard tableaux of the same shape. As a biproduct, we see that

$$ n! = \sum_{\mu \vdash n} (f^\mu)^2, \tag{2.6} $$

since $(f^\mu)^2$ counts the number of pairs of standard tableaux of shape μ.
The *inverse* u^{-1} of a lexicographic word of biletters

$$
u = \begin{pmatrix}
b_1 & b_2 & \cdots & b_k \\
a_1 & a_2 & \cdots & a_k
\end{pmatrix}
$$

is the biletter lexicographic word

$$
u^{-1} = \mathrm{lex} \begin{pmatrix}
a_1 & a_2 & \cdots & a_k \\
b_1 & b_2 & \cdots & b_k
\end{pmatrix},
$$

with "lex(v)" standing for the increasing lexicographic reordering of v. Thus, for

$$
u = \begin{pmatrix}
1 & 1 & 1 & 2 & 2 & 3 & 3 & 4 & 4 & 4 \\
3 & 4 & 5 & 1 & 4 & 1 & 4 & 1 & 2 & 2
\end{pmatrix}
$$

we have

$$u^{-1} = \begin{pmatrix} 1 & 1 & 1 & 2 & 2 & 3 & 4 & 4 & 4 & 5 \\ 2 & 3 & 4 & 4 & 4 & 1 & 1 & 2 & 3 & 1 \end{pmatrix}.$$

Observe that the notion of inverse for a biletter word coincides with the usual notion of inverse in the case of permutations. One can show that in general,

$$u \longleftrightarrow (P, Q) \quad \text{iff} \quad u^{-1} \longleftrightarrow (Q, P). \tag{2.7}$$

Using an approach due to Viennot, in Section 2.5 we give a proof that property (2.7) holds in the case of permutations. The more general version can be shown by an adaptation of this argument. Observing that (2.7) implies that $P(\sigma) = Q(\sigma)$ if and only if σ is an involution, we conclude that the total number $\sum_{\mu \vdash n} f^\mu$ of n-cell standard tableaux is equal to the number of involutions in $\mathfrak{S}_n$.

2.5   Viennot's Shadows

To show that property (2.7) holds for permutations, let us consider the shadow approach described in [Viennot 77]. The idea here is to record the *bumping history* of each cell as the RSK algorithm unfolds. For a cell $(\ell - 1, j - 1)$ in the final common shape, this history takes the form $C_{\ell j} := \{(a_1, b_1), (a_2, b_2), \ldots, (a_k, b_k)\}$, with $a_1 < a_2 < \cdots < a_k$, and is to be understood in the following manner. A pair (a, b) appears in the history of a cell if at step a the value b bumps some previous value b' of this cell. In particular, this means that the bumped value b' will have been inserted, during this same step, in the row that lies immediately above the current row. The following shows how to compute histories of cells in a graphical manner.

We have seen in Section 1.4 that the shadow of a subset E of $\mathbb{N} \times \mathbb{N}$ was characterized by the set $\min(E)$ of its minimal elements. Each cell history is going to be obtained as such a set of minimal elements. To do this, we need the global history R_j of a row, which is defined to be $R_j := \bigcup_\ell C_{\ell j}$. It is somewhat surprising that we can directly compute these sets R_j without prior knowledge of the $C_{\ell j}$. In fact, the $C_{\ell j}$ will be deduced from the R_j. The ideal way to present the upcoming construction is through an animation.[2] The written explanation makes this more awkward than necessary, but we will try to illustrate it with the following example. As our running

---

[2]It is amazing to watch Viennot do this.

example, we consider the permutation $\sigma := (3,1,6,10,2,5,8,4,9,7)$, for which we use RSK to compute the tableau pair

$$\begin{pmatrix} \begin{array}{ccccc} 6 & 10 & & & \\ 3 & 5 & 8 & & \\ 1 & 2 & 4 & 7 & 9 \end{array} & , & \begin{array}{ccccc} 8 & 10 & & & \\ 2 & 5 & 6 & & \\ 1 & 3 & 4 & 7 & 9 \end{array} \end{pmatrix}.$$

This will be compared to the history construction. The general outline of the process is as follows:

(1) We start by setting $R_1 := \{(i, \sigma(i)) \mid 1 \le i \le n\}$. Observe that this corresponds to the graph of σ.

(2) The respective history of cells in the current row, say j, are successively computed using R_j and the history of previous cells in this row. Namely, we set

$$C_{\ell j} := \min\left(R_j \setminus \bigcup_{k<\ell} C_{kj}\right).$$

(3) The global history of the next row is then constructed out of the current row cell histories, by letting R_{j+1} consist of the pairs (a_{i+1}, b_i) such that both pairs (a_{i+1}, b_i) and (a_i, b_i) lie in the same cell history $C_{\ell j}$ for line j.

To help the reader follow our description, we have illustrated the first step of the whole process in Figure 2.7. Here, the respective histories of the cells correspond to individual paths, which are to be read starting at the top and following the direction corresponding to the arrow appearing at the end of the path.

First part. We start with $R_1 := \{(i, \sigma(i)) \mid 1 \le i \le n\}$. The history of the first cell of the first row is set to $C_{11} := \min(R_1) = \{(1,3),(2,1)\}$. For the second cell, we then set $C_{21} := \min(R_1 \setminus C_{11}) = \{(3,6),(5,2)\}$, going on using the general rule in (2) to get

$$C_{31} = \{(4,10),(7,8),(8,4)\},$$
$$C_{41} = \{(7,8),(10,7)\},$$
$$C_{51} = \{(9,9)\}.$$

We continue the process of "peeling off" minimal elements until no cells remain. Then we proceed to treat the second row, starting with its global history $R_2 := \{(2,3),(5,6),(6,10),(8,5),(10,8)\}$ obtained from the $C_{\ell 1}$ using general rule (3). This supposes that the pairs (a_i, b_i) appearing in $C_{\ell j}$ are ordered by increasing value of the a_i. Successive pairs are joined by segments that go from (a_i, b_i) to (a_{i+1}, b_i), and then from (a_{i+1}, b_i) to

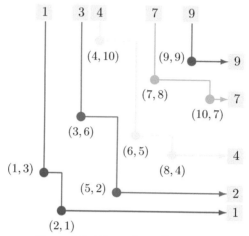

Figure 2.7. History for cells in the first line.

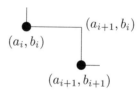

Figure 2.8. Linking points of $C_{\ell j}$.

(a_{i+1}, b_{i+1}), as is shown in Figure 2.8. We then add a segment from (a_1, ∞) to (a_1, b_1) at the beginning, and one from (a_k, b_k) to (∞, b_k) at the end. At this point the first part of the process is over and we obtain the first line $[1, 2, 4, 7, 9]$ of the P-tableau of RSK by reading the labels that appear at the right-hand side of Figure 2.7. We also get the first line $[1, 3, 4, 7, 9]$ of the corresponding Q-tableau by reading the labels that appear at the top. Observing that the graph of σ^{-1} is clearly obtained by reflecting the graph of σ through the $x = y$ line, we deduce that (2.7) holds for first lines of the respective tableaux.

Second part. The second step of the global process consists of starting over with the set of intermediate points (a_{i+1}, b_i) (see Figure 2.9) that have been introduced in the first part. These correspond to values that have been bumped to higher rows.

Remaining parts. We keep on going until we have obtained all rows of the tableaux P and Q.

Exercise. Generalize Viennot's shadow construction to the context of biletter words and pairs of semi-standard tableaux.

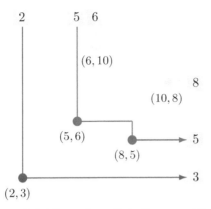

Figure 2.9. History for cells in the second line.

2.6 Charge and Cocharge

Among the most intriguing, but still fundamental, parameters on tableaux are those of charge and cocharge (see [Lascoux and Schützenberger 78]). The original definition makes sense for all skew semi-standard tableaux, but we start with the simpler standard tableau version. Cocharge may best be described in terms of a minimization notion for tableaux. We say that a standard tableau τ has a *reading descent* at cell (i, j) if $\tau(i, j) + 1$ lies in a cell that is to the left of (i, j). For example, the reading descents of the tableau on the left-hand side of Figure 2.10 correspond to the cells that contain the values 2, 4, 7, and 8. The *minimization* $\mathrm{mn}(\tau)$ of a semi-standard tableau τ is the semi-standard tableau whose reading word is the minimization of the reading word of τ. The entries of $\mathrm{mn}(\tau)$ mimic the reading descent pattern of τ: successively reading off the entries of τ from 1 to n, the entries of the corresponding cells in $\mathrm{mn}(\tau)$ stay constant as long as we go eastward, but they go up by one otherwise.

Figure 2.10. Minimization of a tableau.

The *cocharge* $\mathrm{coch}(\tau)$ of a standard tableau τ is the sum of the entries of $\mathrm{mn}(\tau)$. It may appear strange to start with the definition of cocharge rather than with that of charge, but they are directly related to one another and it is easier to start with cocharge. The maximal value for the charged statistic, for a shape λ tableau, corresponds to the standard tableau τ, with $\tau(i,j) = j - 1 + \sum_{k<i} \lambda'_i$. This value is equal to the parameter $n(\lambda)$ defined in Section 1.3. The *charge* $\mathrm{ch}(\tau)$ of a semi-standard tableau τ is just $n(\lambda) - \mathrm{coch}(\tau)$, so that charge is not truly worth an independent description. Many formulas involving the enumeration of tableaux admit a cocharge refinement. For example, we have (see [Macdonald 95])

$$f^\mu(q) := \sum_{\lambda(\tau)=\mu} q^{\mathrm{coch}(\tau)} = q^{n(\mu)} \frac{[n]_q!}{\prod_{c\in\mu}(1 - q^{h(c)})}, \qquad (2.8)$$

where the sum is over the set of standard tableaux of shape μ. We get the right-hand side of the hook length formula (2.1) when we take the limit as t tends to 1 in (2.8). This makes it natural to use the notation $f^\mu(q)$ for this q-analog of the number f^μ of standard tableaux of shape μ.

The cocharge of a standard tableau is the same as the cocharge of its reading word, and we generalize the notion of cocharge to semi-standard tableaux through their reading words. Let w be the reading word of a semi-standard tableau τ. We successively extract subwords $w^{(i)}$ from w until we reach the empty word, and the cocharge of τ is the sum of the cocharges of these words $w^{(i)}$. The first subword $w^{(1)}$ of w is obtained in the following manner. Select the leftmost 1 in w, then the leftmost 2 appearing to the right of this 1, and so on until there is no $k + 1$ to the right of the current value k being considered. At this point, select the leftmost $k + 1$ in w and continue with the previous process until the largest value appearing in w is reached. The subword $w^{(1)}$ thus selected is erased from w and we go on to select $w^{(2)}$ inside the remaining part of w, etc. Borrowing an example of Macdonald, with $w = 322141\ 3$, we get $w^{(1)} = 2143$, $w^{(2)} = 312$ and $w^{(3)} = $. We will encounter later the *Kostka–Foulkes polynomials*

$$K_{\lambda,\mu}(q) = \sum_\tau q^{\mathrm{ch}(\tau)},$$

with sum over the set of semi-standard tableaux of shape λ and content μ, which play an important role. A deeper and more adequate discussion of the charge and cocharge statistics involves the description of the *plactic monoid* and the operation of *cyclage*. For more on this see chapter 6 of [Lothaire 02].

Chapter 3

Invariant Theory

It is now time to start mixing algebraic components into our algebraico-combinatorial recipe. The first of these ingredients comes from invariant theory with some special emphasis on symmetric polynomials. Much of this theory makes systematic use of combinatorial objects, both in the formulation of the main results and in the proof techniques. To learn more about the notions discussed here, refer to [Humphreys 90] and the "bible" of symmetric functions [Macdonald 95].

3.1   The Ring of Polynomials in n Variables

Our overall context is the ring of polynomials, $R = \mathbb{R}[\mathbf{x}]$, in n variables $\mathbf{x} = x_1, x_2, \ldots, x_n$. As introduced in Section 1.2, we use a vectorial notation for monomials, so that polynomials take the form $f(\mathbf{x}) = \sum_{\mathbf{a} \in \mathbb{N}^n} c_{\mathbf{a}} \mathbf{x}^{\mathbf{a}}$ for some choice $c_{\mathbf{a}}$ of coefficients[1], with the sum having finite support (nonzero terms). It is often important to use some monomial order for the terms of a polynomial (see Section 1.4). In such cases we have $f(\mathbf{x}) = c_{\mathbf{a}} \mathbf{x}^{\mathbf{a}} + \cdots$, with omitted terms smaller in monomial order. The monomial $\mathbf{x}^{\mathbf{a}}$ is said to be the *leading monomial* of f, and we write $m(f) := \mathbf{x}^{\mathbf{a}}$. The polynomial f is declared to be *monic* if the coefficient of $m(f)$ in f is equal to 1.

The *degree* $\deg(f)$ of a polynomial f is the maximum degree of monomials that it contains, i.e., $\deg(f) := \max\{\deg(\mathbf{x}^{\mathbf{a}}) \mid c_{\mathbf{a}} \neq 0\}$. The ring R is *graded*[2] with respect to degree. This is to say that there is a natural isomorphism

$$R \simeq \bigoplus_{d \geq 0} R_d, \tag{3.1}$$

---

[1] For the moment we assume $\mathbb{R}$ to be our field of scalars, but most of our statements extend to any field of characteristic zero.

[2] See Section 7.3.

such that $R_d R_k \subseteq R_{d+k}$. Here, the space R_d is the span of all degree d monomials. In a more elementary formulation (3.1) states that polynomials decompose uniquely into *homogeneous components*. To describe this notion more precisely, let us introduce the linear transformation π_d that projects onto the span of monomials of degree d,

$$\pi_d(\mathbf{x}^{\mathbf{a}}) = \begin{cases} \mathbf{x}^{\mathbf{a}} & \text{if } |\mathbf{a}| = d, \\ 0 & \text{otherwise.} \end{cases}$$

Any polynomial $f(\mathbf{x})$ decomposes uniquely in the form $f(\mathbf{x}) = f_0(\mathbf{x}) + f_1(\mathbf{x}) + \cdots + f_d(\mathbf{x})$, with $f_i(\mathbf{x}) := \pi_i\big(f(\mathbf{x})\big)$ corresponding to the ith homogeneous component of $f(\mathbf{x})$. We have already observed that there are $\binom{n+d-1}{n}$ monomials of degree d in n variables, so that the dimension of $R_d = \pi_d(R)$ is

$$\dim R_d = \binom{n+d-1}{d}. \tag{3.2}$$

In general, a subspace $\mathcal{V}$ of R is said to be *homogeneous*, if $\pi_d(\mathcal{V}) \subseteq \mathcal{V}$ for all d. In those cases $\mathcal{V}$ inherits a graded space structure from that of R: $\mathcal{V} \simeq \bigoplus_{d \geq 0} \mathcal{V}_d$, with $\mathcal{V}_d = \pi_d(\mathcal{V})$. In particular, (3.2) implies that

$$\dim \mathcal{V}_d \leq \binom{n+d-1}{d}. \tag{3.3}$$

We can then consider the *Hilbert series*

$$\text{Hilb}_q(\mathcal{V}) := \sum_{d \geq 0} \dim(\mathcal{V}_d) q^d,$$

of $\mathcal{V}$. This formal power series condenses in an efficient format all the information about the respective dimensions of the $\mathcal{V}_d$. The Hilbert series of the space R is

$$\begin{aligned} \text{Hilb}_q(\mathcal{R}) &= \sum_{d \geq 0} \binom{n+d-1}{d} q^d, \\ &= \Big(\frac{1}{1-q}\Big)^n. \end{aligned} \tag{3.4}$$

It is straightforward to check that $\text{Hilb}_q(\mathcal{V} \oplus \mathcal{W}) = \text{Hilb}_q(\mathcal{V}) + \text{Hilb}_q(\mathcal{W})$, and almost as easy to see that $\text{Hilb}_q(\mathcal{V} \otimes \mathcal{W}) = \text{Hilb}_q(\mathcal{V}) \, \text{Hilb}_q(\mathcal{W})$, whenever both sides of this equality make sense. Implicit in this last observation is the fact that there are adequate graded versions of direct sum and tensor product of graded spaces. We will come back to this in Section 7.3, with a somewhat more general point of view.

Scalar Product of Polynomials

It has become customary in our context to denote by ∂x_i the operator of partial derivation with respect to the variable x_i. Any polynomial $f(\mathbf{x})$ gives rise to a partial differential operator obtained by replacing each variable x_i by ∂x_i (derivation with respect to x_i). The resulting operator $f(\partial \mathbf{x})$ allows the introduction of a *scalar product* on R defined by $\langle f(\mathbf{x}), g(\mathbf{x}) \rangle := f(\partial \mathbf{x})g(\mathbf{x})|_{\mathbf{x}=0}$, where we write $\mathbf{x} = 0$ for the simultaneous substitution of 0 for each variable. To make clear that the result is indeed a scalar product, we observe that (for monomials) the definition is equivalent to

$$\langle \mathbf{x}^\mathbf{a}, \mathbf{x}^\mathbf{b} \rangle = \begin{cases} \mathbf{a}! & \text{if } \mathbf{a} = \mathbf{b}, \\ 0 & \text{otherwise,} \end{cases} \tag{3.5}$$

where $\mathbf{a}! = a_1! a_2! \cdots a_n!$. Thus the set of monomials forms an orthogonal basis of R. The reader may easily check that the orthogonal complement of a homogeneous subspace is itself homogeneous.

A surprising feature of the above scalar product stands out when we consider the orthogonal complement of an ideal I. Indeed, we observe that such orthogonal complements coincide with solution sets of systems of partial differential equations:

$$I^\perp = \{g(\mathbf{x}) \mid f(\partial \mathbf{x})g(\mathbf{x}) = 0, \quad f(\mathbf{x}) \in I\}. \tag{3.6}$$

Indeed the inclusion of the right-hand side in $I^\perp$ is evident. To show the reverse inclusion, we observe that for $g(\mathbf{x})$ to be such that $f(\partial \mathbf{x})g(\mathbf{x})|_{\mathbf{x}=0} = 0$ for all $f(\mathbf{x}) \in I$, we must also have[3] $\partial \mathbf{x}^\mathbf{a} f(\partial \mathbf{x})g(\mathbf{x})|_{\mathbf{x}=0} = 0$ for all monomials $\mathbf{x}^\mathbf{a}$, since $\mathbf{x}^\mathbf{a} f(\mathbf{x})$ also lies in the ideal I. Thus follows the apparently much stronger property that $g(\mathbf{x})$ lies in $I^\perp$ if and only if $f(\partial \mathbf{x})g(\mathbf{x}) = 0$ for all $f(\mathbf{x}) \in I$. Equivalently, if $f_1, f_2, \ldots, f_k$ are generators for I, then $g(\mathbf{x}) \in I^\perp$ if and only if

$$\begin{aligned} f_1(\partial x_1, \partial x_2, \ldots, \partial x_n)g(\mathbf{x}) &= 0, \\ f_2(\partial x_1, \partial x_2, \ldots, \partial x_n)g(\mathbf{x}) &= 0, \\ &\vdots \\ f_k(\partial x_1, \partial x_2, \ldots, \partial x_n)g(\mathbf{x}) &= 0. \end{aligned} \tag{3.7}$$

Another easy but useful observation is that $I^\perp$ is closed under derivation, since partial derivatives commute.

---

[3]We extend our vectorial notation to differential monomials, setting $\partial \mathbf{x}^\mathbf{a} := \partial x_1^{a_1} \cdots \partial x_n^{a_n}$.

Group Actions on Polynomials

While our discussion is concentrated on the symmetric group and its action on polynomials by permutation of the variables, much can be stated in a more general setup. Various extensions to these more general situations have been a driving force in recent developments in the field. To promote a broader outlook we start with a more global point of view, but then rapidly specialize back to the symmetric group case so readers may still feel that they are in familiar territory.

Let G be a finite subgroup of the group $O_n(\mathbb{R})$ of orthogonal matrices for the usual dot product

$$v \bullet w := v_1 w_1 + \cdots + v_n w_n$$

on $\mathbb{R}^n$. One considers the natural action of G on polynomials $f(\mathbf{x})$ in $\mathbb{R}[\mathbf{x}]$. More precisely, for γ in G we set $\gamma \cdot f(\mathbf{x}) := f(\mathbf{x}\gamma)$, with $\mathbf{x}\gamma$ standing for the usual product between $\mathbf{x} = (x_1, x_2, \ldots, x_n)$ and the matrix γ. Among the classical examples are the group of permutation matrices and the *hyperoctahedral group* B_n, whose elements are signed permutations. These are the $n \times n$ matrices with exactly one ± 1 entry on each line and each column. The action of B_n on polynomials consists of permuting variables and changing signs of some of them.

For all finite groups of orthogonal matrices, the above action is degree preserving, since variables are sent to linear combinations of variables. We have

$$\langle \gamma \cdot \mathbf{x}^{\mathbf{a}}, \gamma \cdot \mathbf{x}^{\mathbf{b}} \rangle = \langle \mathbf{x}^{\mathbf{a}}, \mathbf{x}^{\mathbf{b}} \rangle \tag{3.8}$$

for all $\gamma \in G$ and monomials $\mathbf{x}^{\mathbf{a}}$ and $\mathbf{x}^{\mathbf{b}}$. To see this, introduce two sets of n commuting variables $\mathbf{y}$ and $\mathbf{z}$. Observe that the monomial $\mathbf{x}^{\mathbf{a}}$ is the coefficient of $\mathbf{y}^{\mathbf{a}}$ in $\exp(\mathbf{x} \bullet \mathbf{y})$, and that Taylor's expansion theorem can be expressed in the form $\exp(\mathbf{y} \bullet \partial \mathbf{x}) f(\mathbf{x}) = f(\mathbf{x} + \mathbf{y})$. For the purpose of what follows, the G-action is extended in a straightforward manner to functions and operators. We can now show[4] globally that (3.8) holds just by proving the single identity

$$\gamma \cdot \exp(\mathbf{y} \bullet \partial \mathbf{x}) \exp(\mathbf{z} \bullet \mathbf{x}) = \exp\big(\mathbf{y} \bullet (\gamma \cdot \partial \mathbf{x})\big) \exp\big(\mathbf{z} \bullet (\gamma \cdot \mathbf{x})\big). \tag{3.9}$$

We first expand the left-hand side using Taylor's expansion to get

$$\gamma \cdot \exp(\mathbf{y} \bullet \partial \mathbf{x}) \exp(\mathbf{z} \bullet \mathbf{x}) = \gamma \cdot \exp\big(\mathbf{z} \bullet (\mathbf{x} + \mathbf{y})\big)$$
$$= \exp(\mathbf{z} \bullet \mathbf{y}) \exp\big(\mathbf{z} \bullet (\gamma \cdot \mathbf{x})\big).$$

---

[4]Following a proof by A. Garsia [Garsia and Haiman 08].

Then, using the fact that γ is orthogonal, we check that the right-hand side of (3.9) expands to

$$
\begin{aligned}
\exp\big(\mathbf{y} \bullet (\gamma \cdot \partial \mathbf{x})\big) \exp\big(\mathbf{z} \bullet (\gamma \cdot \mathbf{x})\big) &= \exp\big((\gamma^{-1} \cdot \mathbf{y}) \bullet \partial \mathbf{x}\big) \exp\big((\gamma^{-1} \cdot \mathbf{z}) \bullet \mathbf{x}\big) \\
&= \exp\big((\gamma^{-1} \cdot \mathbf{z}) \bullet (\mathbf{x} + \gamma^{-1} \cdot \mathbf{y})\big) \\
&= \exp\big((\gamma^{-1} \cdot \mathbf{z}) \bullet (\gamma^{-1} \cdot \mathbf{y})\big) \exp\big((\gamma^{-1} \cdot \mathbf{z}) \bullet \mathbf{x}\big) \\
&= \exp(\mathbf{z} \bullet \mathbf{y}) \exp\big(\mathbf{z} \bullet (\gamma \cdot \mathbf{x})\big),
\end{aligned}
$$

thus verifying equality in (3.9). We are particularly interested in *G-invariant subspaces* $\mathcal{V}$ of R, namely those for which $\gamma \cdot \mathcal{V} \subseteq \mathcal{V}$ for all γ in G. Whenever $\mathcal{V}$ is homogeneous, in addition to being G-invariant, each of its homogeneous components $\mathcal{V}_d$ is also clearly a G-invariant subspace.

Reflection Groups

An important special case corresponds to groups generated by reflections. Recall that a *reflection* is a linear transformation of finite order which fixes a hyperplane pointwise. It is well known (see Figure 3.1) that for a hyperplane $H_v := \{x \in \mathbb{R}^n \mid v \bullet \mathbf{x} = 0\}$, the *reflection* s_v into H_v affords the simple formula $s_v(w) := w - 2\frac{w \bullet v}{v \bullet v} w$. We say that H_v is a *reflecting hyperplane* of G, if s_v is in G. For a group generated by reflections to be finite, strong conditions on angles between the reflecting hyperplanes of G must be met. This is illustrated in Figure 3.2, with the group of symmetries of a polyhedron sometimes called the *truncated cuboctahedron*. When trying to discuss these angle conditions further we are naturally led to translate everything in terms of families of vectors orthogonal to the reflecting hyperplanes. These vectors are the "roots" discussed in the next section.

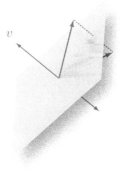

Figure 3.1. Hyperplane reflection.

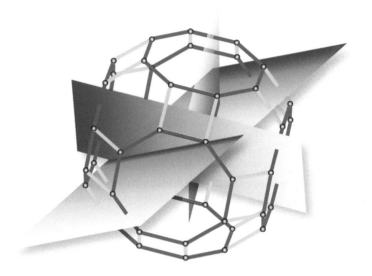

Figure 3.2. Angles between reflecting hyperplanes.

3.2 Root Systems

Given a finite group G generated by reflections, we consider the set Φ of vectors α such that $s_\alpha \in G$ and $|\alpha| = 1$, called *roots*, and call Φ the *root system* associated with G. Observe that Φ is a finite set such that $-\alpha \in \Phi$ if $\alpha \in \Phi$. Moreover, $\gamma \cdot \alpha \in \Phi$ whenever $\gamma \in G$ and $\alpha \in \Phi$. One says that $\alpha^\vee := 2\alpha/(\alpha \bullet \alpha)$ is a *coroot*. The kernel of the natural homomorphism of G into the group of permutations of Φ is clearly trivial. The notion of root system plays a central role in the classification of simple Lie algebras, and the groups G considered appear as the associated Weyl groups.

To have a natural means of systematically choosing one vector out of each of the pairs $\alpha, -\alpha$ we choose a total ordering on $\mathbb{R}^n$, setting $\alpha < \beta$ if and only if the first nonzero component of $\beta - \alpha$ is positive. We can then define Φ^+ to be the set of *positive roots* of G, i.e., $\alpha \in \Phi$ with $\alpha > 0$. Thus the set of reflecting hyperplanes of G is the set of hyperplanes having an equation of the form $\alpha \bullet \mathbf{x} = 0$, with $\alpha \in \Phi^+$. The group G acts as a reflection group on the linear span $\mathcal{V}$ of the root system. The associated *weights* are the the vectors λ in $\mathcal{V}$ such that $\lambda \bullet \alpha^\vee$ is an integer for all roots α. The set Λ of weights forms a lattice, and the set λ^+ of *positive weights* is

$$\lambda^+ := \{\lambda \in \Lambda \mid \lambda \bullet \alpha \geq 0, \alpha \in \Phi\}.$$

To understand the reason for considering these notions, let us work out their meaning for the symmetric group $G = S_n$. The roots are the differences $\alpha_{ij} := e_i - e_j$, with $e_i \in \mathbb{R}_n$ denoting the unit vector having a 1 in its ith coordinate. Positive roots are those α_{ij} for which $i < j$. The symmetric group acts as a reflection group on the span $\mathcal{V}$ of the roots. This is the subspace of $\mathbb{R}^n$ of vectors having coordinate sum equal to 0. Coroots and roots are identified in this case. The roots $e_i - e_{i+1}$ are called *simple roots*. The weights correspond to vectors with integer entries, and the positive weights are the partitions having lengths at most $n - 1$.

All these notions are key in the generalization of Macdonald polynomials (see Section 9), but these generalizations will not be further discussed here. For those interested in learning more, consult [Macdonald 03].

3.3   Coxeter Groups

Among the interesting reflection groups the Coxeter groups are certainly worth further attention. We recall that a (finite) *Coxeter system* is a pair (W, S), where W is a finite group and S is a set of generators for W subject to *Coxeter relations* of the form $(st)^{m_{st}} = 1$, for $s, t \in S$. Here the m_{st} are positive integers such that $m_{ss} = 1$, and $m_{st} = m_{ts} \geq 2$, for $s \neq t$. Often the generator set S is implicit and we only mention the *Coxeter group* W. The Coxeter relations can be encoded in the form of a *Coxeter diagram* whose nodes are the elements of S. We put a (labeled) edge between two nodes when $m_{st} \geq 3$. For $m_{st} = 3$ we draw a simple unlabeled edge, for $m_{st} = 4$ we draw a double edge, and for m_{st} with larger values we label the edge with this value.

Writing $S_1 + S_2$ for the disjoint union[5] of the sets S_1 and S_2, it is easy to check that the product $(W_1 \times W_2, S_1 + S_2)$ of two Coxeter systems is also a Coxeter system. In this case, every element s in S_1 commutes with each t in S_2. Thus, in those cases $m_{st} = 2$, implying that in the Coxeter diagram there are no edges between the vertices in S_1 and those in S_2. The classification of all finite Coxeter groups reduces to the classification of *irreducible* Coxeter groups, meaning those that are not isomorphic to products of two nontrivial Coxeter groups. In particular, the Coxeter diagram of an irreducible Coxeter group must have only one connected component. The classification has been carried out, and the only possible irreducible Coxeter groups are those given in Figure 3.3. The subscripts correspond to the number of elements in the corresponding generator sets. Groups of type $\mathbf{A}_n$ are the classical symmetric groups $\mathfrak{S}_{n+1}$, the generator set being the set of transpositions $(i, i + 1)$, $1 \leq i \leq n$.

---

[5]This will be our convention throughout.

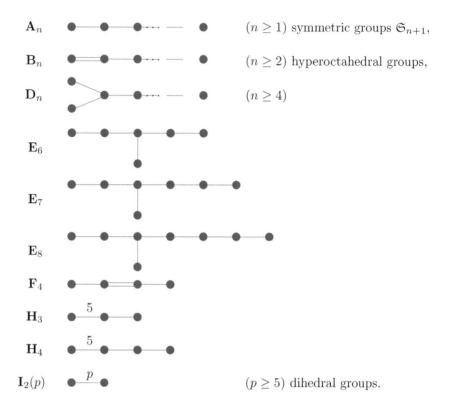

Figure 3.3. Irreducible Coxeter groups.

3.4   Invariant and Skew-Invariant Polynomials

Let us go back to our discussion of special subspaces of the ring of polynomials, with respect to the action of G. Our first step along this road is
to consider the notions of invariant and skew-invariant polynomials for G.
These generalize respectively the notions of symmetric and antisymmetric
polynomials.

It is classical in invariant theory to denote by R^G the homogeneous subspace of G-*invariant polynomials* consisting of the polynomials $f(\mathbf{x})$ such
that $\gamma \cdot f(\mathbf{x}) = f(\mathbf{x})$ for all γ in G. Here, not only is the subspace R^G
invariant, but each of its individual elements is. For the symmetric group,
invariant polynomials are just symmetric polynomials. For the hyperoctahedral group, they correspond to symmetric polynomials in the squares
of the variables. In the case of groups generated by reflections, it is well
known (see [Humphreys 90]) that R^G is in fact a subring of R for which

we can find generator sets of n homogeneous algebraically independent elements $f_1, \ldots, f_n$, whose respective degrees will be denoted by $d_1, \ldots, d_n$. This is to say that any G-invariant polynomial can be uniquely expanded as a polynomial in the f_i. For example, for the symmetric group, the polynomials f_i may be chosen to be the *power sum*

$$p_i(\mathbf{x}) = x_1^i + x_2^i + \cdots + x_n^i,$$

with i going from 1 to n. In Section 3.5 we will come back to how to find the unique expansion of a given symmetric polynomial in terms of these f_i. For the hyperoctahedral group, we can choose the polynomials $p_{2i}(\mathbf{x})$ with $1 \leq i \leq n$.

Although the f_i are not uniquely characterized, the d_i are basic numerical invariants of the group called the *degrees of G*. Any n-set $\{f_1, \ldots, f_n\}$ of invariants with these properties is called a *set of basic invariants* for G. It follows that the Hilbert series of R^G takes the form

$$\mathrm{Hilb}_q(R^G) = \prod_{i=1}^{n} \frac{1}{1 - q^{d_i}}, \tag{3.10}$$

with the d_i standing for the degrees of G. To easily construct invariant polynomials, we make use of *Reynold's symmetrization operator* which maps $f(\mathbf{x})$ to $\rho_G\big(f(\mathbf{x})\big) := \frac{1}{|G|} \sum_{\gamma \in G} \gamma \cdot f(\mathbf{x})$. This linear map, which projects R onto R^G, makes it easy to explicitly construct a spanning set for all homogeneous G-invariant polynomials of a given degree d. Indeed, we simply apply ρ_G to each degree d monomial $\mathbf{x}^{\mathbf{a}}$.

Skew-Invariant Polynomials

At the other end of the spectrum is the notion of skew-invariant polynomial which generalizes that of antisymmetric polynomial. Recall first that our groups are generated by reflections (or pseudo-reflections if we work over the complex numbers[6]). It follows that $\det(\gamma) = \pm 1$ for all elements of G. Furthermore, this is a natural generalization for the notion of sign of a permutation. With this in mind, a polynomial is said to be *skew invariant* with respect to G if $\gamma \cdot f(\mathbf{x}) = \det(\gamma) f(\mathbf{x})$, for all γ in G. Just as in the invariant case, we have a *Reynold skewing operator*,

$$f(\mathbf{x}) \mapsto \frac{1}{|G|} \sum_{\gamma \in G} \det(\gamma) \gamma \cdot f(\mathbf{x}),$$

---

[6]See [Kane 01] for more on this natural expansion. In this case, we should replace $\det(\gamma)$ by $\det(\gamma)^{-1}$ in the definition of skew-invariant.

mapping R onto the subspace of skew-invariants. Indeed, this can be checked by the following simple computation:

$$\kappa \cdot \frac{1}{|G|} \sum_{\gamma \in G} \det(\gamma) \gamma \cdot f(\mathbf{x}) = \det(\kappa) \frac{1}{|G|} \sum_{\gamma \in G} \det(\kappa) \det(\gamma)(\kappa\gamma) \cdot f(\mathbf{x})$$

$$= \det(\kappa) \frac{1}{|G|} \sum_{\gamma \in G} \det(\gamma) \gamma \cdot f(\mathbf{x}).$$

Moreover, if $f(\mathbf{x})$ is skew-invariant, we clearly have

$$\frac{1}{|G|} \sum_{\gamma \in G} \det(\gamma) \gamma \cdot f(\mathbf{x}) = \frac{1}{|G|} \sum_{\gamma \in G} \det(\gamma) \det(\gamma) f(\mathbf{x})$$

$$= f(\mathbf{x}).$$

If $f(\mathbf{x})$ is an invariant polynomial and $g(\mathbf{x})$ is skew-invariant, then the product $f(\mathbf{x})g(\mathbf{x})$ is skew-invariant. This turns the subspace $R^{\pm}$ of skew-invariant polynomials into a graded R^G-module. Just as was discussed for G-invariants, we can use a Reynold skewing operator to explicitly construct a spanning set for all homogeneous skew-invariant polynomials of degree d. Observe that the result of the skewing operator may very well be 0.

A particularly interesting skew-invariant polynomial is given by the *Jacobian determinant*,

$$\Delta_G(\mathbf{x}) := \det \frac{\partial f_i}{\partial x_j}, \qquad (3.11)$$

associated with any specific set $\{f_1, \ldots, f_n\}$ of basic invariants for G. It is striking that the resulting polynomial (up to a scalar multiple) does not depend on the actual choice of the f_i (see [Humphreys 90, Section 3.13] for an elegant proof), but it is even more important for our discussion that the Jacobian determinant is the minimal degree (unique up to a scalar multiple) G-skew-invariant polynomial. The more precise statement is that $g(\mathbf{x})$ is skew-invariant if and only if it can be written as $g(\mathbf{x}) = f(\mathbf{x})\Delta_G(\mathbf{x})$, with $f(\mathbf{x})$ being G-invariant.

Let us check by direct computation that $\Delta_G(\mathbf{x})$ is skew-invariant. Recall that the f_i are G-invariant, so that we need only consider the effect of γ on the derivation variables when computing as follows:

$$\gamma \cdot \Delta_G(\mathbf{x}) = \Delta_G(\gamma\mathbf{x}) = \det\left(\gamma \cdot \frac{\partial f_i}{\partial x_j}\right) = \det(\gamma) \det\left(\frac{\partial f_i}{\partial x_j}\right).$$

In view of (3.10) and (3.11), the above observation implies that the Hilbert series of the homogeneous invariant subspace $R^{\pm}$ is just

$$\mathrm{Hilb}_q(R^{\pm}) = \prod_{i=1}^{n} \frac{q^{d_i} - 1}{1 - q^{d_i}}, \tag{3.12}$$

with the d_i standing for the degrees of G, just as in (3.10). We will elaborate on these notions in the special case of the symmetric group in Section 3.9.

3.5 Symmetric Polynomials and Functions

We have already stressed that the case of the symmetric group $\mathfrak{S}_n$ plays a special role in our global discussion. Let us thus expand for this particular case the notions that have just been outlined.

The $\mathfrak{S}_n$-invariant polynomials correspond to *symmetric polynomials*. Recall that these are the polynomials $f(\mathbf{x})$ such that, for each permutation σ of the set $\{1, 2, \ldots, n\}$, we have

$$f(x_{\sigma(1)}, x_{\sigma(2)}, \ldots, x_{\sigma(n)}) = f(x_1, x_2, \ldots, x_n).$$

As in the more general context of reflection groups acting on polynomials, the graded ring $R^{\mathfrak{S}_n}$ decomposes as a direct sum $\bigoplus_{d=0}^{\infty} R_d^{\mathfrak{S}_n}$, with $R_d^{\mathfrak{S}_n}$ standing for the homogeneous degree d component of $R^{\mathfrak{S}_n}$. The dimension of $R_d^{\mathfrak{S}_n}$ is the number of partitions of d with at most n parts.

A linear basis of $R_d^{\mathfrak{S}_n}$ is given by the *monomial symmetric polynomials* defined forthwith. We write $m_\lambda = m_\lambda(\mathbf{x})$ for the sum of all the different monomials $\mathbf{x}^{\mathbf{a}}$ for which $\widehat{\mathbf{a}} = \lambda$. In other words, the sum is over all the monomials $\mathbf{x}^{\mathbf{a}}$ with $\mathbf{a}$ varying in the set of rearrangements of the length n vector $(\lambda_1, \ldots, \lambda_k, 0, \ldots, 0)$. This makes it clear that a given monomial appears with multiplicity one in m_λ, as illustrated by the fact that

$$m_{211}(x_1, x_2, x_3) = x_1^2 x_2 x_3 + x_1 x_2^2 x_3 + x_1 x_2 x_3^2$$

contains three terms rather than six. Observe that our definition forces $m_\lambda = 0$ when $\ell(\lambda) > n$.

The linear independence of the m_λ, for λ varying in the set of partitions of d with length at most n, follows by an obvious triangularity argument from that of the monomials $\mathbf{x}^{\lambda} = x_1^{\lambda_1} x_2^{\lambda_2} \cdots x_k^{\lambda_k}$. Indeed, one version of the fundamental theorem of algebra states precisely that these are linearly independent. The fact that the set $\{m_\lambda \mid \lambda \vdash d\}$ spans $R_d^{\mathfrak{S}_n}$ is a direct consequence of the definition of symmetric polynomial.

Further examples of monomial symmetric polynomials, for $\mathbf{x} = x_1, x_2, x_3, x_4$, are

$$m_5 = x_1^5 + x_2^5 + x_3^5 + x_4^5,$$

$$m_{41} = x_1^4 x_2 + x_1^4 x_3 + x_1^4 x_4 + x_2^4 x_3 + x_2^4 x_4 + x_3^4 x_4$$
$$+ x_1 x_2^4 + x_1 x_3^4 + x_1 x_4^4 + x_2 x_3^4 + x_2 x_4^4 + x_3 x_4^4,$$

$$m_{32} = x_1^3 x_2^2 + x_1^3 x_3^2 + x_1^3 x_4^2 + x_2^3 x_3^2 + x_2^3 x_4^2 + x_3^3 x_4^2$$
$$+ x_1^2 x_2^3 + x_1^2 x_3^3 + x_1^2 x_4^3 + x_2^2 x_3^3 + x_2^2 x_4^3 + x_3^2 x_4^3,$$

$$m_{311} = x_1^3 x_2 x_3 + x_1^3 x_2 x_4 + x_1^3 x_3 x_4 + x_2^3 x_3 x_4$$
$$+ x_1 x_2^3 x_3 + x_1 x_2^3 x_4 + x_1 x_3^3 x_4 + x_2 x_3^3 x_4$$
$$+ x_1 x_2 x_3^3 + x_1 x_2 x_4^3 + x_1 x_3 x_4^3 + x_2 x_3 x_4^3,$$

$$m_{221} = x_1^2 x_2^2 x_3 + x_1^2 x_2^2 x_4 + x_1^2 x_3^2 x_4 + x_2^2 x_3^2 x_4$$
$$+ x_1^2 x_2 x_3^2 + x_1^2 x_2 x_4^2 + x_1^2 x_3 x_4^2 + x_2^2 x_3 x_4^2$$
$$+ x_1 x_2^2 x_3^2 + x_1 x_2^2 x_4^2 + x_1 x_3^2 x_4^2 + x_2 x_3^2 x_4^2,$$

$$m_{2111} = x_1^2 x_2 x_3 x_4 + x_1 x_2^2 x_3 x_4 + x_1 x_2 x_3^2 x_4 + x_1 x_2 x_3 x_4^2$$

$$m_{11111} = 0.$$

Complete Homogeneous Symmetric Polynomials

A second classical set of symmetric polynomials consists of the *complete homogeneous symmetric polynomials*, so called because they correspond to the sum of all monomials of a given degree: $h_d = h_d(\mathbf{x}) = \sum_{|\mathbf{a}|=d} \mathbf{x}^{\mathbf{a}}$. As we have just done, it is customary to omit mention of the actual variables $\mathbf{x}$ in symmetric polynomial expressions. Most formulas that we will obtain are in fact independent of the actual set of variables. For example, we have

$$h_d = \sum_{\lambda \vdash d} m_\lambda \tag{3.13}$$

which gives the expression of h_d in the basis of monomial symmetric functions. We may also specify the h_d through their generating function

$$H(\zeta) := \sum_{d \geq 0} h_d \zeta^d = \prod_{i=1}^{n} \frac{1}{1 - x_i \zeta}. \tag{3.14}$$

Observe that this requires that $h_0 = 1$. We extend the notion of complete homogeneous symmetric polynomials to partitions λ, setting $h_\lambda := h_{\lambda_1} h_{\lambda_2} \cdots h_{\lambda_r}$, for $\lambda = \lambda_1 \lambda_2 \ldots \lambda_r$. Clearly h_λ is homogeneous of degree

$d = |\lambda|$. Expanding the complete homogeneous symmetric functions in terms of the monomial basis, we get

$$h_{11111} = m_5 + 5m_{41} + 10m_{32} + 20m_{311} + 30m_{221} + 60m_{2111} + 120m_{11111},$$
$$h_{2111} = m_5 + 4m_{41} + 7m_{32} + 13m_{31,1} + 18m_{221} + 33m_{2111} + 60m_{11111},$$
$$h_{221} = m_5 + 3m_{41} + 5m_{32} + 8m_{311} + 11m_{221} + 18m_{2111} + 30m_{11111},$$
$$h_{311} = m_5 + 3m_{41} + 4m_{32} + 7m_{311} + 8m_{221} + 13m_{2111} + 20m_{11111},$$
$$h_{32} = m_5 + 2m_{41} + 3m_{32} + 4m_{311} + 5m_{221} + 7m_{2111} + 10m_{11111},$$
$$h_{41} = m_5 + 2m_{41} + 2m_{32} + 3m_{311} + 3m_{221} + 4m_{2111} + 5m_{11111},$$
$$h_5 = m_5 + m_{41} + m_{32} + m_{311} + m_{221} + m_{2111} + m_{11111}.$$

Elementary Symmetric Polynomials

In a similar manner, *elementary symmetric polynomials* e_k are introduced via their generating series

$$E(\zeta) := \sum_{k \geq 0} e_k \zeta^k = \prod_{i=1}^{n} (1 + x_i \zeta). \tag{3.15}$$

Observe that e_k is just another name for $m_{(1^n)}$. The interesting identity

$$H(\zeta)E(-\zeta) = 1 \tag{3.16}$$

can easily be deduced from (3.14) and (3.15). Comparing coefficients of ζ^n on both sides of this equation, we get

$$\sum_{k=0}^{n} (-1)^k h_{n-k} e_k = 0 \tag{3.17}$$

for $n > 0$. As before, we set $e_\lambda := e_{\lambda_1} e_{\lambda_2} \cdots e_{\lambda_r}$. Observe that e_λ is zero whenever λ has a part that is larger than n, the number of variables. In terms of the monomial basis, the elementary polynomials expand as

$$e_{11111} = m_5 + 5m_{41} + 10m_{32} + 20m_{311} + 30m_{221} + 60m_{2111} + 120m_{11111},$$
$$e_{2111} = m_{41} + 3m_{32} + 7m_{311} + 12m_{221} + 27m_{2111} + 60m_{11111},$$
$$e_{221} = m_{32} + 2m_{311} + 5m_{221} + 12m_{2111} + 30m_{11111},$$
$$e_{311} = m_{311} + 2m_{221} + 7m_{2111} + 20m_{11111},$$
$$e_{32} = m_{221} + 3m_{2111} + 10m_{11111},$$
$$e_{41} = m_{2111} + 5m_{11111},$$
$$e_5 = m_{11111}.$$

Power Sums

Another classical family is that of the *power sums*: $p_k = p_k(\mathbf{x}) := x_1^k + \cdots + x_n^k$, for $k \geq 1$, with p_λ equal to $p_{\lambda_1} p_{\lambda_2} \cdots p_{\lambda_r}$. The following computation is entirely straightforward:

$$
\begin{aligned}
H(\zeta) &= \exp\left(\sum_{i=1}^n \log\left(\frac{1}{1 - x_i \zeta}\right)\right) \\
&= \exp\left(\sum_{i=1}^n \sum_{k \geq 1} x_i^k \frac{\zeta^k}{k}\right) \\
&= \exp\left(P(\zeta)\right),
\end{aligned}
\tag{3.18}
$$

where $P(\zeta) := \sum_{k \geq 1} p_k \frac{\zeta^k}{k}$. Expanding back the right-hand side as a power series in ζ and comparing like powers of ζ on both sides, we obtain

$$
h_d = \sum_{\mu \vdash d} \frac{1}{z_\mu} p_\mu,
\tag{3.19}
$$

with z_μ as defined in Section 1.6. A similar computation, starting with (3.15) gives

$$
e_d = \sum_{\mu \vdash d} \frac{(-1)^{d - \ell(\mu)}}{z_\mu} p_\mu.
\tag{3.20}
$$

Recall that $e_d = 0$ when d is larger than n. It follows that in those cases equation (3.20) gives an algebraic relation between the p_k. For instance, with $n = 2$, we have

$$
e_3(x_1, x_2) = \frac{1}{6} p_1(x_1, x_2)^3 - \frac{1}{2} p_1(x_1, x_2) p_2(x_1, x_2) + \frac{1}{3} p_3(x_1, x_2) = 0.
$$

To express the p_μ in the monomial basis, we introduce the notion of coarsening of a partition. Namely, for μ and ν both partitions of the same integer n (μ being of length k) we say that $\varphi \colon \{1, \ldots, k\} \to \mathbb{N}$ is a ν-*coarsening* of μ if $\sum_{\varphi(j)=i} \mu_j = \nu_i$ for all i going from 1 to the length of ν. The relevant result can then be expressed as follows:

$$
p_\mu = \sum_\nu L_{\mu\nu} m_\nu,
\tag{3.21}
$$

with $L_{\mu\nu}$ equal to the number of ν-coarsenings of μ. This follows easily from expanding out the product p_μ (see [Macdonald 95, p. 103]).

To illustrate, we have

$$p_{11111} = m_5 + 5m_{41} + 10m_{32} + 20m_{311} + 30m_{221} + 60m_{2111} + 120m_{11111},$$
$$p_{2111} = m_5 + 3m_{41} + 4m_{32} + 6m_{311} + 6m_{221} + 6m_{2111},$$
$$p_{221} = m_5 + m_{41} + 2m_{32} + 2m_{221},$$
$$p_{311} = m_5 + 2m_{41} + m_{32} + 2m_{311},$$
$$p_{32} = m_5 + m_{32},$$
$$p_{41} = m_5 + m_{41},$$
$$p_5 = m_5.$$

Infinite Number of Variables

All of these families make sense when n goes to infinity, and we use the term *symmetric functions* for the corresponding elements of the graded inductive limit $R^{\mathfrak{S}}$, of the rings $R^{\mathfrak{S}_n}$. We naturally denote by $R_d^{\mathfrak{S}}$ the degree d homogeneous component of this inductive limit. Working in $R^{\mathfrak{S}}$ has the advantage of removing special conditions related to the number of variables. The relevant fact here is that both the families $\{e_k\}_{k \geq 1}$ and $\{h_k\}_{k \geq 1}$ are algebraically independent when the number of variables becomes infinite (denumerable). Moreover, all identities obtained in $R^{\mathfrak{S}}$ remain valid when we restrict the underlying set of variables to a finite subset of these variables.

One way to become more familiar with this approach is to consider a direct generalization of the familiar binomial coefficient formula. Consider the monomial expansion of the powers of $h_1 = x_1 + x_2 + \cdots$:

$$h_1{}^2 = m_2 + 2m_{11},$$
$$h_1{}^3 = m_3 + 3m_{21} + 6m_{111},$$
$$h_1{}^4 = m_4 + 4m_{31} + 6m_{22} + 12m_{211} + 24m_{1111},$$
$$h_1{}^5 = m_5 + 5m_{41} + 10m_{32} + 20m_{311} + 30m_{221} + 60m_{2111} + 120m_{11111}.$$

If we restrict the number of variables to two, then any monomial m_μ vanishes if μ has more than two parts, and the identities above become

$$h_1{}^2 = m_2 + 2m_{11},$$
$$h_1{}^3 = m_3 + 3m_{21},$$
$$h_1{}^4 = m_4 + 4m_{31} + 6m_{22},$$
$$h_1{}^5 = m_5 + 5m_{41} + 10m_{32}.$$

These are just rewritings of the usual binomial expansion formulas. Passing to a denumerable set of variables is just a simple way of keeping all the significant terms, and there are only a finite number of them.

3.6 The Fundamental Theorem of Symmetric Functions

The fundamental result concerning symmetric functions (which goes back at least to Newton) is the following.

Theorem 3.1. *Every symmetric function can be written as a polynomial in the elementary symmetric functions e_k, $k = 1, 2, 3, \ldots$.*

Proof: To prove the theorem[7] we make use of $(0,1)$-matrices $A \colon \mathbb{N} \times \mathbb{N} \longrightarrow \{0,1\}$ with finitely many nonzero entries. Let us denote by $\mathrm{col}(A)$ the *column-sum vector* of A and by $\mathrm{row}(A)$ its *row-sum vector*. Then let $\lambda(A)$ stand for the partition obtained by sorting the nonzero entries of $\mathrm{row}(A)$ in decreasing order. Observe that $\lambda(A) < \nu$ (in lexicographic order), whenever $\mathrm{col}(A) = \nu'$. The matrix $A = (a_{ij})$ is said to be *left justified* if no zero appears to the left of a one in $\mathrm{row}(A)$. Let us define the *matrix monomial* $\mathbf{x}^A$ to be $\prod_{i,j \geq 1} x_i^{a_{ij}}$. In other words, the exponent of x_i in $\mathbf{x}^A$ is the sum of the integers that appear on row i in A. Clearly, the indices $i_1, i_2, \ldots, i_k$ of the lines where 1 appears in a given column of A can be bijectively encoded by the monomial $x_{i_1} x_{i_2} \cdots x_{i_k}$. Through this encoding, the elementary symmetric function $e_k(\mathbf{x})$ corresponds to all possible choices of positions of k ones in a column. It follows immediately that $e_{\nu'}(\mathbf{x}) = \sum_{\mathrm{col}(A)=\nu'} \mathbf{x}^A$, where the sum runs over the set of left-justified matrices satisfying the condition $\mathrm{col}(A) = \nu'$. The extra condition that $\lambda(A) = \nu$ forces each line itself to be left justified, hence it follows that

$$m_\nu(\mathbf{x}) = \sum_{\substack{\mathrm{col}(A)=\nu' \\ \lambda(A)=\nu'}} \mathbf{x}^A. \tag{3.22}$$

Thus, we get the formula $e_{\nu'} = m_\nu + \sum_{\mu < \nu} M_{\nu,\mu} m_\mu$, where $M_{\nu,\mu}$ is the number of $(0,1)$-matrices A such that $\mathrm{col}(A) = \nu'$ and $\mathrm{row}(A) = \mu$. The punch line here is that the transformation from the $e_{\nu'}$ to the m_μ is given by an upper triangular matrix with diagonal entries all equal to 1. We conclude that the set $\{e_\nu\}_{\nu \vdash d}$ is a basis of $R_d^{\mathfrak{S}}$, since we already know that the m_ν form a basis. $\square$

3.7 More Basic Identities

Using the generating functions $H(\zeta)$, $E(\zeta)$ and $P(\zeta)$, we can easily derive more basic identities between the families considered in the previous section. One of these comes from computing the derivative of (3.18) to get

---

[7]This is heavily inspired by Stanley's presentation, see [Stanley 97, Section 7.4].

$H'(\zeta) = P'(\zeta)H(\zeta)$. Comparing coefficients of ζ^d on both sides, we deduce that $dh_d = \sum_{r=1}^{d} p_r h_{d-r}$. We then use Cramer's rule to solve this system of equations for the p_k, and find the following expansion of the p_k in terms of the h_λ:

$$p_k = (-1)^{k-1} \det \begin{vmatrix} h_1 & 1 & 0 & \cdots & 0 \\ 2h_2 & h_1 & 1 & \cdots & 0 \\ 3h_3 & h_2 & h_1 & \cdots & 0 \\ \vdots & \vdots & \vdots & \ddots & \vdots \\ kh_k & h_{k-1} & h_{k-2} & \cdots & h_1 \end{vmatrix}. \tag{3.23}$$

Similarly, from (3.15), we get $de_d = \sum_{r=1}^{d}(-1)^{r-1}p_r e_{d-r}$, from which we derive

$$p_k = \det \begin{vmatrix} e_1 & 1 & 0 & \cdots & 0 \\ 2e_2 & e_1 & 1 & \cdots & 0 \\ 3e_3 & e_2 & e_1 & \cdots & 0 \\ \vdots & \vdots & \vdots & \ddots & \vdots \\ ke_k & e_{k-1} & e_{k-2} & \cdots & e_1 \end{vmatrix}. \tag{3.24}$$

We can recursively solve (3.17) either to write the h_k in terms of e_j, or vice-versa. Considering all of the relations that we have found thus far, we conclude the following.

Proposition 3.2. *The ring $R^{\mathfrak{S}_n}$ of symmetric polynomials in n variables is isomorphic to the ring $\mathbb{R}[e_1, e_2, \ldots, e_n]$ of polynomials in the e_i. Moreover, letting n go to infinity, it is also the case that $R^{\mathfrak{S}} = \mathbb{R}[p_1, p_2, \ldots] = \mathbb{R}[h_1, h_2, \ldots]$.*

The ω Involution

Many of the notions and formulas involving symmetric functions come in pairs that are naturally tied together through a natural linear and multiplicative involution, denoted by ω. It is probably easiest to define it in terms of the power sum, setting $\omega(p_k) := (-1)^{k-1}p_k$. It immediately follows, say comparing (3.19) and (3.20), that $\omega(h_k) = e_k$. In particular, this makes it clear that ω translates (3.23) into (3.24), and vice-versa. Observe that $\omega(p_\mu) = \varepsilon(\mu)p_\mu$, with $\varepsilon(\mu) = (-1)^{|\mu|-\ell(\mu)}$ equal to the sign of any permutation having cycle structure equal to μ.

Explicit Expansions for Small n

Just to feel more comfortable with all this, let us give the explicit expansions of the various bases in terms of one another for small values of n.

First, we evidently have $m_1 = e_1 = h_1 = p_1$, and the various expansions of the monomial basis for $n = 2$ are

$$m_2 = p_2 = e_1^2 - 2e_2 = 2h_2 - h_1^2,$$

$$m_{11} = e_2 = h_1^2 - h_2 = \frac{1}{2}(p_1^2 - p_2).$$

The rest of the $n = 2$ expansions are covered by the identities

$$p_1^2 = e_1^2 = h_1^2 = m_2 + m_{11},$$

$$h_2 = m_2 + m_{11} = e_1^2 - e_2 = \frac{1}{2}(p_1^2 + p_2).$$

For the $n = 3$ case, a more systematic presentation is given by the following table, where all expressions along a given row are equal.

m_3	$e_1^3 - 3e_1e_2 + 3e_3$	$h_1^3 - 3h_1h_2 + 3h_3$	p_3
m_{21}	$e_1e_2 - 3e_3$	$-2h_1^3 + 5h_1h_2 - 3h_3$	$p_1p_2 - p_3$
m_{111}	e_3	$h_1^3 - 2h_1h_2$	$\frac{1}{6}(p_1^3 - 3p_1p_2 + 2p_3)$
$m_{21} + 3m_{111}$	e_1e_2	$h_1^3 - h_1h_2$	$\frac{1}{2}(p_1^3 - p_1p_2)$
$m_3 + 3m_{21} + 6m_{111}$	e_1^3	h_1^3	p_1^3
$m_3 + m_{21} + m_{111}$	$e_1^3 - 2e_1e_2 + e_3$	h_3	$\frac{1}{6}(p_1^3 + 3p_1p_2 + 2p_3)$
$m_3 + 2m_{21} + 3m_{111}$	$e_1^3 - e_1e_2$	h_1h_2	$\frac{1}{2}(p_1^3 + p_1p_2)$
$m_3 + m_{21}$	$e_1^3 - 2e_1e_2$	$-h_1^3 + 2h_1h_2$	p_1p_2

To further explore the various identities and relations between symmetric functions, the reader should refer to J. Stembridge's Maple package SF [Stembridge 09]. He has to be thanked for most of the explicit calculations that appear in this book.

3.8   Plethystic Substitutions

Many formulas and manipulations of symmetric functions become a lot clearer if we use notations and operations coming from the study of λ-rings (see [Knutson 73]) and plethystic operations. Plethysm operations were first defined in [Littlewood 50] (see also [Macdonald 95]). In this approach, we are encouraged to think of symmetric functions as abstract operations on expressions involving some set of variables, on which the effect of a simple power sum is specified. Typically p_k is the operation that raises all the variables to the power k. This is naturally extended to all symmetric functions as described below.

Suppose that A is a rational fraction, $A \in \mathbb{R}(\mathbf{z})$, in some set of variables $\mathbf{z} = z_1, z_2, \ldots$. We set $p_k[A] := A\big|_{z_i \mapsto z_i^k}$, which is to say that each z_i

is replaced by z_i^k in A. Clearly, we have $p_k[A+B] = p_k[A] + p_k[B]$ and $p_k[AB] = p_k[A]p_k[B]$. In particular, writing $\mathbf{z} = z_1 + z_2 + \cdots$, we observe that $p_k[\mathbf{z}] = z_1^k + z_2^k + \cdots$, so that $p_k[\mathbf{z}]$ is just the usual power sum in the variables $\mathbf{z}$. We extend the above operation to any symmetric function $f = \sum_\lambda a_\lambda p_\lambda$ (here expanded in terms of the power sum basis) by setting

$$f[A] := \sum_\lambda a_\lambda \prod_{i=1}^{\ell(\lambda)} p_{\lambda_i}[A]. \qquad (3.25)$$

We then say that this is a *plethystic substitution* of A in f. Observe that this definition implicitly supposes that we have $(f+g)[A] = f[A] + g[A]$, and $(f \cdot g)[A] = f[A]g[A]$. It follows in particular that

$$\sum_{n \geq 0} h_n[A]\zeta^n = \exp\left(\sum_{k \geq 1} p_k[A]\frac{\zeta^k}{k}\right). \qquad (3.26)$$

It is often easier to calculate a plethystic substitution using the formal series $H(\zeta)$, $E(\zeta)$ and $P(\zeta)$, denoting these substitutions by

$$H[A;\zeta] := \sum_{k \geq 0} h_k[A]\zeta^k, \qquad E[A;\zeta] := \sum_{k \geq 0} e_k[A]\zeta^k,$$

$$P[A;\zeta] := \sum_{k \geq 1} p_k[A]\frac{\zeta^k}{k}.$$

A word of caution may be in order here. In general, plethystic substitutions do not commute with other operations. In particular, they do not commute with evaluation. Forgetting this can lead to obviously erroneous identities.

The Plethysm $f[x]$

The simplest case corresponds to the substitution of a single variable x in a symmetric function, so that p_k becomes x^k. Using (3.18) and (3.15), we immediately deduce that $h_n[x] = x^n$ and that $e_n[x] = 0$ for all $n \geq 2$. We also have $m_\mu[x] = 0$, whenever μ has more than one part. It is also the case that $h_n[1] = 1$ for all n, and that $e_n[1] = 0$ for all $n \geq 2$. However, we have

$$h_n[-1] = \begin{cases} 1 & \text{if } n = 0, \\ -1 & \text{if } n = 1, \\ 0, & \text{otherwise,} \end{cases} \quad \text{and} \quad e_n[-1] = \begin{cases} 1 & \text{if } n \text{ is even,} \\ -1 & \text{otherwise.} \end{cases}$$

Indeed, using $p_k[-1] = -1$, we calculate that

$$H[-1; \zeta] = \exp\left(\sum_{k \geq 1} -\frac{\zeta^k}{k}\right) = 1 - \zeta,$$

$$E[-1; \zeta] = \exp\left(\sum_{k \geq 1}(-1)^k \frac{\zeta^k}{k}\right) = \frac{1}{1 + \zeta}.$$

The Plethysm $f[-\varepsilon z]$

Let us introduce a formal variable with the property that $\varepsilon^d = (-1)^d$, so that $f[\varepsilon z] = (-1)^d f(z)$ for any homogeneous degree d symmetric function $f(z)$. At first glance, this may appear a bit strange, but it is nevertheless interesting. Indeed, we verify immediately that $p_k[-\varepsilon z] = (-1)^{k-1} p_k(z)$. Hence, $f[-\varepsilon z]$ coincides with $\omega(f(z))$. Observe also that, with f still homogeneous of degree d, we have $f[-z] = (-1)^d \omega(f(z))$, so that $e_n[-z] = (-1)^n h_n(z)$.

The Plethysm $f[1 - u]$

Since $p_k[1 - u] = 1 - u^k$, we calculate that

$$\sum_{n \geq 0} h_n[1 - u]\zeta^n = \exp\left(\sum_{k \geq 1}(1 - u^k)\frac{\zeta^k}{k}\right)$$

$$= \exp\left(\sum_{k \geq 1}\frac{\zeta^k}{k}\right)\exp\left(-\sum_{k \geq 1}\frac{(u\zeta)^k}{k}\right) \qquad (3.27)$$

$$= \frac{1 - u\zeta}{1 - \zeta} = 1 + \sum_{k \geq 1}(1 - u)\zeta^k.$$

Comparing both sides of the equality, we conclude that $h_k[1 - u] = 1 - u$ if $k > 0$. We will see later (see Section 4.5) how to use this to compute the corresponding plethystic substitution $s_\mu[1 - u]$ into the "Schur functions" (introduced in Chapter 4).

The Plethysm $e_n[(1 - t)(1 - q)]$

Just as above we calculate that

$$\sum_{n \geq 1} e_n[(1 - t)(1 - q)]\zeta^n = \frac{\zeta(1 - q)(1 - t)}{(1 + t\zeta)(1 + q\zeta)}.$$

In particular, this implies that

$$\frac{(-1)^{n-1}}{(1 - t)(1 - q)}e_n[(1 - t)(1 - q)] = [n]_{q,t}, \qquad (3.28)$$

using the notation

$$[n]_{q,t} = \frac{q^n - t^n}{q - t}$$
$$= q^{n-1} + q^{n-2}t + \cdots + qt^{n-2} + t^{n-1}.$$

The Plethysm $h_n[\mathbf{z}/(1-t)]$

As a last example, let us compute $h_n[\mathbf{z}/(1-t)]$. We start with (3.19) to derive

$$h_n\left[\frac{\mathbf{z}}{1-t}\right] = \sum_{\mu \vdash n} \frac{1}{z_\mu} \prod_{i=1}^{\ell(\mu)} \frac{p_{\mu_i}(\mathbf{z})}{1-t^{\mu_i}}. \tag{3.29}$$

For example, we have

$$h_1\left[\frac{\mathbf{z}}{1-t}\right] = \frac{h_1}{1-t}$$

$$h_2\left[\frac{\mathbf{z}}{1-t}\right] = \frac{t\,h_1^2 + (1-t)\,h_2}{(1-t)\,(1-t^2)}$$

$$h_3\left[\frac{\mathbf{z}}{1-t}\right] = \frac{t^3\,h_1^3 + t(1-t)(1+2\,t)h_1\,h_2 + (1-t)(1-t^2)h_3}{(1-t)\,(1-t^2)\,(1-t^3)}.$$

Exercise. Find explicit expressions for $h_n[1/(1-t)]$ and $e_n[1/(1-t)]$.

3.9 Antisymmetric Polynomials

Specializing the definition of skew-invariant (see Section 3.4) to the case of $\mathfrak{S}_n$, we get the notion of *antisymmetric polynomial*. More explicitly, these are the polynomials such that $\sigma \cdot P(\mathbf{x}) = \mathrm{sgn}(\sigma)P\mathbf{x})$ for all σ in $\mathfrak{S}_n$. A typical example corresponds to the *Vandermonde determinant*:

$$\Delta_n(\mathbf{x}) := \sum_{\sigma \in \mathfrak{S}_n} \mathrm{sgn}(\sigma)\sigma \cdot \mathbf{x}^{\delta_n},$$

$$= \det \begin{vmatrix} x_1^{n-1} & x_1^{n-2} & \cdots & 1 \\ x_2^{n-1} & x_2^{n-2} & \cdots & 1 \\ \vdots & \vdots & \ddots & \vdots \\ x_n^{n-1} & x_n^{n-2} & \cdots & 1 \end{vmatrix} \tag{3.30}$$

with $\delta_n := (n-1, n-2, \ldots, 2, 1, 0)$. These are the $\mathfrak{S}_n$ versions of the Jacobian determinant. Indeed, we get (3.30) as a special case of definition (3.10) by choosing f_i equal to $p_i(\mathbf{x})/i$. To highlight the fundamental role that $\Delta_n(\mathbf{x})$ plays in the study of antisymmetric polynomials, we specialize a general (see Section 3.4) property of Jacobian determinants to the current context.

Proposition 3.3. *Any antisymmetric polynomial can be written as the product of the Vandermonde determinant by a symmetric polynomial. Moreover, the Vandermonde determinant factorizes as*

$$\Delta_n(\mathbf{x}) = \prod_{i<j}(x_i - x_j).$$

Proof: To see this, we first observe that any antisymmetric polynomial $f(\mathbf{x})$ vanishes whenever two of its variables are equal. In particular, consider the transposition $\tau = (i, j)$ that exchanges the two variables x_i and x_j. The sign of τ being -1, we get $\tau \cdot f(\mathbf{x}) = -f(\mathbf{x})$ from the antisymmetry of $f(\mathbf{x})$. On the other hand, when $x_i = x_j$ we clearly have $\tau \cdot f(\mathbf{x}) = f(\mathbf{x})$, hence $f(\mathbf{x})|_{x_i = x_j} = 0$. It follows that for all $i < j$ the linear polynomial $x_i - x_j$ divides $f(\mathbf{x})$. Hence the polynomial $\pi_n(\mathbf{x}) := \prod_{i<j}(x_i - x_j)$ divides $f(\mathbf{x})$. The resulting quotient $f(\mathbf{x})/\pi_n(\mathbf{x})$ is a clearly a symmetric polynomial, since

$$\sigma \cdot \frac{f(\mathbf{x})}{\pi_n(\mathbf{x})} = \frac{\sigma \cdot f(\mathbf{x})}{\sigma \cdot \pi_n(\mathbf{x})} = \frac{\varepsilon(\sigma)f(\mathbf{x})}{\varepsilon(\sigma)\pi_n(\mathbf{x})} = \frac{f(\mathbf{x})}{\pi_n(\mathbf{x})}.$$

To see that $\pi_n(\mathbf{x})$ actually coincides with $\Delta_n(\mathbf{x})$, we need only observe that the two polynomials have the same degree, hence their quotient must be constant. One then checks that this constant is 1, since both polynomials have $x_1^{n-1}x_2^{n-2}\cdots x_{n-1}$ as leading coefficients.   □

For any n-vector $\mathbf{a} = (a_1, a_2, \ldots, a_n) \in \mathbb{N}^n$, with all entries distinct, the determinant[8]

$$\Delta_\mathbf{a}(\mathbf{x}) := \det \begin{vmatrix} x_1^{a_1} & x_1^{a_2} & \cdots & x_1^{a_n} \\ x_2^{a_1} & x_2^{a_2} & \cdots & x_2^{a_n} \\ \vdots & \vdots & \ddots & \vdots \\ x_n^{a_1} & x_n^{a_2} & \cdots & x_n^{a_n} \end{vmatrix}$$

is evidently antisymmetric, since a permutation of the variables corresponds to a permutation of the rows in the matrix. Up to a column permutation, we might as well suppose that $a_1 > a_2 > \cdots > a_n \geq 0$, since this only changes the sign of the determinant. Any such vector $\mathbf{a}$ can be written in the form $\mathbf{a} = \lambda + \delta_n$ for some partition λ. Observe also that the case $\mathbf{a} = \delta_n$ corresponds exactly to the Vandermonde determinant $\Delta_n(\mathbf{x})$. In view of our previous discussion, the quotient $\Delta_{\lambda+\delta}(\mathbf{x})/\Delta_n(\mathbf{x})$ is a symmetric polynomial. In a sense that will be made clear in the sequel, these are the most fundamental symmetric polynomials. Although apparently[9] first

---

[8]Observe that $\Delta_\mathbf{a}(\mathbf{x})$ appears as a special case (see Section 1.3) of determinants associated with diagrams $\mathbf{d}$. In this case all cells have second coordinates equal to zero.

[9]See [Macdonald 95, p. 61].

introduced by C. Jacobi in [Jacobi 41], they have come to be known as *Schur polynomials*. As shown by I. Schur (see [Schur 01]), they naturally arise as characters of irreducible polynomial representations of $O_n(\mathbb{R})$. The *symmetric polynomials* s_λ are defined by

$$s_\lambda(\mathbf{x}) := \Delta_{\lambda+\delta_n}(\mathbf{x})/\Delta_n(\mathbf{x}). \qquad (3.31)$$

As we will see later, the set of Schur polynomials s_λ constitutes a linear basis of the ring of symmetric functions.

It may be worth mentioning at this point that the Jacobian determinant of a reflection group affords a linear factor decomposition that is very similar to the factorization of the Vandermonde determinant described in Proposition 3.3. In this factorization there are as many linear factors as there are reflections in the group. Each of these factors is of the form $v \bullet \mathbf{x}$, with v running over the set of positive roots associated with the group.

Exercise. For any family $A = \big(\alpha_j(\zeta)\big)_{1 \le j \le n}$ of n formal power series

$$\alpha_j(\zeta) = \sum_{k \ge 0} a_{j,k}\zeta^k,$$

consider the antisymmetric function

$$\Delta_A(\mathbf{x}) := \det\big(\alpha_j(x_i)\big)_{1 \le i,j \le n}.$$

Compute the coefficient of s_λ in the Schur basis expansion of the symmetric function $s_A(\mathbf{x}) := \Delta_A(\mathbf{x})/\Delta_n(\mathbf{x})$.

Chapter 4

Schur Functions and Quasisymmetric Functions

Schur polynomials play a crucial role in a wide variety of mathematical contexts. This may be related to the fact that they have several very different natural descriptions. We begin our survey of their properties by an account of the most important of these descriptions. We also discuss how basic identities can be derived from these various points of view. Since one of these points of view involves the enumeration of semi-standard tableaux of skew shape, it also seems natural to discuss the quasisymmetric functions that appear when we enumerate semi-standard fillings of diagrams. Along the way, we will review the basis of the theory of poset partitions introduced by R. Stanley, and how quasisymmetric functions arise in this context. For more on poset partitions and quasisymmetric functions, we refer the reader to [Stanley 97].

4.1 A Combinatorial Approach

For a partition λ (here considered as a diagram), the Schur polynomial $s_\lambda(\mathbf{x})$ in the n variables $\mathbf{x} = x_1, \ldots, x_n$ affords the combinatorial description[1]

$$s_\lambda(\mathbf{x}) := \sum_\tau \mathbf{x}_\tau, \qquad (4.1)$$

with the sum being over all semi-standard tableaux $\tau \colon \lambda \to \{1, 2, \ldots, n\}$, and $\mathbf{x}_\tau := \prod_{c \in \lambda} x_{\tau(c)}$ being the *evaluation monomial* of τ. This definition

---

[1]This can be shown to coincide with definition (3.31).

$$
\begin{array}{cccc}
b & b & c & b \\
a \quad a & a \quad b & a \quad b & a \quad c
\end{array}
$$

Figure 4.1. Types of semi-standard 21-shape tableaux.

is naturally extended to skew shapes λ/μ to get *skew Schur* polynomials,

$$s_{\lambda/\mu}(\mathbf{x}) := \sum_\tau \mathbf{x}_\tau, \tag{4.2}$$

with the sum over semi-standard tableaux of shape λ/μ. To illustrate, let $a < b < c$ and consider the possible semi-standard tableaux of shape 21 as shown in Figure 4.1. This shows that the Schur function s_{21} expands as $m_{21} + 2m_{111}$.

Exercise. Verify directly from (4.2) that $s_{4321/321} = h_1^4$, and generalize.

If we assume (as is shown in the exercise below) that this definition results in $s_\lambda(\mathbf{x})$ being a symmetric polynomial, then equation (4.1) implies that

$$s_\lambda(\mathbf{x}) = \sum_{\mu \vdash n} K_{\lambda,\mu} m_\mu(\mathbf{x}). \tag{4.3}$$

Thus, from (2.2) we get

$$
\begin{aligned}
s_4 &= m_4 + m_{31} + m_{22} + m_{211} + m_{1111}, \\
s_{31} &= m_{31} + m_{22} + 2m_{211} + 3m_{1111}, \\
s_{22} &= m_{22} + m_{211} + 2m_{1111}, \\
s_{211} &= m_{211} + 3m_{1111}, \\
s_{1111} &= m_{1111},
\end{aligned}
$$

and generally $s_{(n)} = \sum_{\mu \vdash n} m_\mu = h_n$, and $s_{1^n} = m_{1^n} = e_n$.

Exercise. Extend equation (4.3) to the context of skew Schur functions. Prove that $s_{\lambda/\mu}(\mathbf{x})$ is invariant under the exchange of the variables x_i and x_{i+1}, by constructing for each λ/μ-shape semi-standard tableau τ a new semi-standard tableau of the same shape but in which the respective number of occurrences of i and $i+1$ are exchanged.

For many discussions it is more natural to think that a tableau is filled with the variables x_i, rather than their indices i, with order $x_1 < x_2 < x_3 < \cdots$. With this point of view, the evaluation monomial is just the product of all entries of the tableau. We can now easily deduce from (4.1) a "summation formula" for Schur functions. One considers $s_\lambda(\mathbf{x} + \mathbf{y})$ as the evaluation of the Schur function s_λ on the *alphabet of variables*

$$\mathbf{x} + \mathbf{y} = x_1 + x_2 + x_3 + \cdots + y_1 + y_2 + y_3 + \cdots,$$

with the convention that the y_k are larger than the x_i. We find that

$$s_\lambda(\mathbf{x} + \mathbf{y}) = \sum_{\mu \subseteq \lambda} s_\mu(\mathbf{x}) s_{\lambda/\mu}(\mathbf{y}), \qquad (4.4)$$

since any semi-standard filling $\tau : \lambda \to \{x_1, x_2, x_3, \ldots, y_1, y_2, y_3, \ldots\}$ can be naturally separated into the subpartition $\mu \subseteq \lambda$ where only the x_i appear, and the skew shape λ/μ where only the y_k appear. Considering the special cases $\lambda = (n)$ and $\lambda = 1^n$, we get

$$h_n(\mathbf{x} + \mathbf{y}) = \sum_{k=0}^{n} h_k(\mathbf{x}) h_{n-k}(\mathbf{y}) \qquad (4.5)$$

and

$$e_n(\mathbf{x} + \mathbf{y}) = \sum_{k=0}^{n} e_k(\mathbf{x}) e_{n-k}(\mathbf{y}). \qquad (4.6)$$

It follows from Section 3.8 that

$$h_n[\mathbf{x} - \mathbf{y}] = \sum_{k=0}^{n} (-1)^{n-k} h_k(\mathbf{x}) e_{n-k}(\mathbf{y}).$$

4.2 Formulas Derived from Tableau Combinatorics

Exploiting the Robinson–Schensted–Knuth correspondence, we find more interesting formulas. Let us first consider the correspondence between length k lexicographic biletter words on the alphabet $\mathbb{N}^+ \times \mathbb{N}^+$ and pairs of semi-standard tableaux of the same shape λ (a partition of k). For such a biletter word

$$\binom{\mathbf{b}}{\mathbf{a}} = \begin{pmatrix} b_1 & b_2 & \cdots & b_k \\ a_1 & a_2 & \cdots & a_k \end{pmatrix},$$

define the associated evaluation monomial to be

$$\mathbf{x_a y_b} := x_{a_1} x_{a_2} \cdots x_{a_n} y_{b_1} y_{b_2} \cdots y_{b_n}.$$

Observe that this establishes a bijection $\binom{\mathbf{b}}{\mathbf{a}} \leftrightarrow \mathbf{x_a y_b}$ between lexicographic biletter words and pairs of monomials, respectively in the variables $\mathbf{x}$ and $\mathbf{y}$. We may identify the set of length k lexicographic words with terms of the sum $\sum_{\mathbf{a},\mathbf{b}} \mathbf{x_a y_b}$, with $\mathbf{a}$ and $\mathbf{b}$ in $(\mathbb{N}^+)^k$. In turn, this sum is readily seen to be the same as the plethystic evaluation $h_k[\mathbf{xy}]$ of h_k at $\mathbf{xy} = \sum_{i,j\geq 1} x_i y_j$. Recall that by definition we have

$$\sum_{k\geq 0} h_k[\mathbf{xy}] = \sum_{i,j\geq 1} \frac{1}{1 - x_i y_j}.$$

We now apply the RSK correspondence to each monomial $\mathbf{x_a y_b}$ (translated back into a lexicographic biletter word) to get a pair of same shape semi-standard tableaux (P, Q) for which we clearly have $\mathbf{x_a y_b} = \mathbf{x}_P \mathbf{y}_Q$. Hence, we can rewrite the sum $\sum_{\mathbf{a},\mathbf{b}} \mathbf{x_a y_b}$ as $\sum_{P,Q} \mathbf{x}_P \mathbf{y}_Q$, with the sum now being over the pairs of semi-standard tableaux of the same shape. It follows that we have the identity

$$h_k(\mathbf{xy}) = \sum_{\lambda \vdash k} s_\lambda(\mathbf{x}) s_\lambda(\mathbf{y}). \tag{4.7}$$

It is easy to deduce (applying ω) that we also have

$$e_k(\mathbf{xy}) = \sum_{\lambda \vdash k} s_{\lambda'}(\mathbf{x}) s_\lambda(\mathbf{y}). \tag{4.8}$$

Now, summing both sides of (4.7) for all n in $\mathbb{N}$, we get the Cauchy–Littlewood formula:

$$\prod_{i,j\geq 1} \frac{1}{1 - x_i y_j} = \sum_\lambda s_\lambda(\mathbf{x}) s_\lambda(\mathbf{y}). \tag{4.9}$$

Let us restrict this argument to words of the form $u = \left(\begin{smallmatrix} 1 & 2 & \cdots & k \\ a_1 & a_2 & \cdots & a_k \end{smallmatrix} \right)$ while setting the evaluation monomial of u be $\mathbf{x_a}$. To make the correspondence compatible, we also set the evaluation monomial of a pair (P, Q) to be $\mathbf{x}_P$. We then get the identity

$$(x_1 + x_2 + x_3 + \cdots)^n = h_1^n = \sum_{\lambda \vdash n} f^\lambda s_\lambda(\mathbf{x}), \tag{4.10}$$

with f^λ standing for the number of standard tableaux of shape λ. For example, we have

$$h_1^2 = s_2 + s_{11},$$
$$h_1^3 = s_3 + 2s_{21} + s_{111},$$
$$h_1^4 = s_4 + 3s_{31} + 2s_{22} + 3s_{211} + s_{1111},$$
$$h_1^5 = s_5 + 4s_{41} + 5s_{32} + 6s_{311} + 5s_{221} + 4s_{2111} + s_{11111}.$$

It is interesting to note that by expanding both sides of (4.10) in terms of the basis p_μ and taking the coefficient of $p_1^n/n!$, we get back identity (5.4).

4.3 Dual Basis and Cauchy Kernel

Our next description of Schur functions is essentially a recipe for a recursive construction. We introduce a scalar product on $R^{\mathfrak{S}}$ by the simple device of defining its effect on a specific linear basis, that of the p_μ. Indeed we set

$$\langle p_\lambda, p_\mu \rangle := \begin{cases} z_\lambda & \text{if } \lambda = \mu, \\ 0 & \text{if } \lambda \neq \mu. \end{cases} \tag{4.11}$$

Observe that it follows from the definitions that ω preserves the scalar product, i.e., $\langle \omega(f), \omega(g) \rangle = \langle f, g \rangle$. For example, we calculate directly that

$$\langle h_n, h_n \rangle = \langle e_n, e_n \rangle = \sum_{\mu \vdash n} \frac{1}{z_\mu}.$$

We will see below that this evaluates to 1.

The Schur polynomials can be obtained through a Gramm–Schmidt orthogonalization process applied to the basis of monomial symmetric $\{m_\lambda\}_{\lambda \vdash n}$, written in increasing lexicographic order of partitions. More precisely, the Schur polynomials are uniquely determined by the following two properties:

$$(1) \quad \langle s_\lambda, s_\mu \rangle = 0, \quad \text{whenever } \lambda \neq \mu,$$

$$(2) \quad s_\lambda = m_\lambda + \sum_{\mu \prec \lambda} c_{\lambda\mu} m_\mu. \tag{4.12}$$

This last statement is actually stronger that what is announced above. More precisely, we obtain the same Schur function s_λ using any linear extension of the partial order of dominance.

On the other hand we have the following simple computation:

$$\begin{aligned}
\prod_{i,j \geq 1} \frac{1}{1 - x_i y_j} &= \exp \sum_{i,j} -\log(1 - x_i y_j) \\
&= \exp \sum_{i,j} \sum_{k \geq 1} (x_i y_j)^k / k \\
&= \exp \sum_{k \geq 1} p_k(\mathbf{x}) p_k(\mathbf{y}) / k \\
&= \sum_\lambda p_\lambda(\mathbf{x}) \frac{p_\lambda(\mathbf{y})}{z_\lambda}.
\end{aligned} \tag{4.13}$$

The left-hand side of (4.9) and (4.13) is called the *Cauchy kernel*. It is henceforth denoted by $\Omega(\mathbf{xy})$. One can directly check that

$$\Omega(\mathbf{xy}) = \sum_\lambda h_\lambda(\mathbf{x}) m_\lambda(\mathbf{y}). \tag{4.14}$$

Various similar identities characterize *dual basis pairs* $\{u_\lambda\}_\lambda$ and $\{v_\mu\}_\mu$. These are pairs of bases such that

$$\langle u_\lambda, v_\mu \rangle = \begin{cases} 1 & \text{if } \lambda = \mu, \\ 0 & \text{otherwise.} \end{cases} \tag{4.15}$$

The relevant statement here is that conditions (4.15) are globally equivalent to the single identity

$$\Omega(\mathbf{xy}) = \sum_\lambda u_\lambda(\mathbf{x}) v_\lambda(\mathbf{y}). \tag{4.16}$$

Indeed, to show the equivalence of (4.15) and (4.16), we respectively expand p_λ and p_μ/z_μ in the given bases u_λ and v_μ. Supposing that these expansions are $p_\lambda = \sum_\rho a_{\lambda\rho} v_\rho$, and $p_\mu/z_\mu = \sum_\nu b_{\mu\nu} v_\nu$, then by definition (4.11) of scalar product we must have $\delta_{\lambda,\mu} = \langle p_\lambda, p_\mu/z_\mu \rangle = \sum_{\rho,\nu} a_{\rho\lambda} \langle u_\lambda, v_\nu \rangle b_{\nu\mu}$. In matrix form, this becomes $AZB^{\mathrm{Tr}} = \mathrm{Id}$, where we have set $A := (a_{\lambda\rho})$, $B := (b_{\mu\nu})$, and $Z := (\langle u_\lambda, v_\mu \rangle)$. Statement (4.15) is clearly equivalent to $Z = \mathrm{Id}$, and thus it is also equivalent to $AB^{\mathrm{Tr}} = \mathrm{Id}$. On the other hand, in view of (4.13), identity (4.16) corresponds to

$$\sum_\lambda u_\lambda(\mathbf{x}) v_\lambda(\mathbf{y}) = \sum_\lambda p_\lambda(\mathbf{x}) \frac{p_\lambda(\mathbf{y})}{z_\lambda}.$$

In other words,

$$\sum_\lambda \left(\sum_\rho a_{\lambda\rho} p_\rho(\mathbf{x}) \right) \left(\sum_\nu b_{\lambda\nu} \frac{p_\nu(\mathbf{y})}{z_\nu} \right) = \sum_\lambda p_\lambda(\mathbf{x}) \frac{p_\lambda(\mathbf{y})}{z_\lambda}.$$

We then observe that the $p_\lambda(\mathbf{x}) p_\lambda(\mathbf{y})/z_\lambda$ are linearly independent, hence

$$\sum_\lambda a_{\lambda\rho} b_{\lambda\nu} = \delta_{\rho,\nu}$$

is equivalent to (4.16). Going back to matrix form, this is exactly the statement that $A^{\mathrm{Tr}} B = \mathrm{Id}$, which is now evidently equivalent to (4.15).

In particular, identities (4.9) and (4.15) are equivalent to $\langle s_\lambda, s_\mu \rangle = \delta_{\lambda,\mu}$ and $\langle m_\lambda, h_\mu \rangle = \delta_{\lambda,\mu}$, where $\delta_{\lambda,\mu}$ is the well-known[2] *Kronecker delta function*. In other words, $\{s_\lambda\}_\lambda$ is a self dual orthonormal basis. Exploiting the duality between $\{m_\lambda\}_\lambda$ and $\{h_\mu\}_\mu$, and using equality (4.3), we get

$$h_\mu = \sum_\lambda K_{\lambda,\mu} s_\lambda. \tag{4.17}$$

---

[2]Which is so "well known" that we always seem to be required to mention the fact.

Comparing this with equation (4.3), we notice that the index of summation is now λ rather than μ. This will be reformulated in terms of transposition of transition matrices in Section 4.4.

There is, however, one glaring piece missing, which is the family of *forgotten symmetric functions*, denoted f_μ by Macdonald. This is just the dual basis for the e_μ. As we will see in Section 10.6, they do play an active role. For example, we have

$$f_4 = -s_4 + s_{31} - s_{211} + s_{1111},$$
$$f_{31} = 2s_4 - s_{31} - s_{22} + s_{211},$$
$$f_{22} = s_4 - s_{31} + s_{22},$$
$$f_{211} = -3s_4 + s_{31},$$
$$f_{1111} = s_4.$$

The Kronecker Product

A second important product on symmetric functions, arising in the study of Schur–Weyl duality, is the *Kronecker product* also called the *internal product*. This commutative and associative bilinear product is most easily defined in terms of the power sum basis by setting

$$p_\lambda * p_\mu := \begin{cases} z_\lambda p_\lambda & \text{if } \lambda = \mu, \\ 0 & \text{otherwise.} \end{cases} \tag{4.18}$$

One readily observes that $h_n * f = f$ for all degree n homogeneous symmetric functions f. Likewise we see that $e_n * f = \omega(f)$. Using associativity and commutativity of the Kronecker product as well as this last identity, we easily check that $\omega(f) * g = f * \omega(g)$. The problem of computing the multiplicities of irreducible representations in the tensor product of two irreducible representations corresponds to computing the structure constants of the Kronecker product, i.e., the $\gamma_{\mu\nu}^\lambda$ such that $s_\mu * s_\nu = \sum_\lambda \gamma_{\mu\nu}^\lambda s_\lambda$. For example,

$$s_{321} * s_{321} = s_6 + 2s_{51} + 3s_{42} + 4s_{411} + 2s_{33} + 5s_{321} + 4s_{3111}$$
$$+ 2s_{222} + 3s_{2211} + 2s_{21111} + s_{111111}.$$

4.4   Transition Matrices

We now describe the matrices that express the possible changes of basis between the six fundamental bases that have been discussed. For this, let us borrow Figure 4.2 from [Macdonald 95, p. 104]. The oriented edges are labeled by transition matrices between bases. The relevant bases are

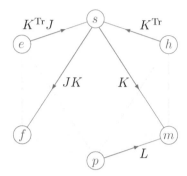

Figure 4.2. Transition matrices.

indicated by labels on end vertices. Thus, the arrow labeled $L = (L_{\mu\nu})$, from vertex p to vertex m, stands for the matrix whose entries correspond to equation (3.21). In a similar manner, K stands for the Kostka matrix, and J for the matrix that corresponds to the involution ω in the Schur basis. As usual we write K^{Tr} for the transpose of K. All of the unlabeled edges of the graph in Figure 4.2 can be oriented (in any which way) and filled in so that the complete diagram commutes. For the sake of clarity, we have omitted diagonal edges (such as the one that goes from s to p, corresponding essentially to the character table of $\mathfrak{S}_n$).

Exercise. Compute the character table of $\mathfrak{S}_4$ via an explicit calculation of $L^{-1}K$. See Section 5.6 for the answer.

4.5   Jacobi–Trudi Determinants

The *Jacobi–Trudi formula* gives an explicit expansion of Schur functions in terms of complete homogeneous functions. Namely, we have

$$s_\mu = \det(h_{\mu_i + j - i})_{1 \le i, j \le n}, \qquad (4.19)$$

with $h_k = 0$ whenever $k < 0$. Equivalently, in terms of elementary symmetric functions, we have the *dual Jacobi–Trudi formula*

$$s_{\mu'} = \det(e_{\mu_i + j - i})_{1 \le i, j \le n}, \qquad (4.20)$$

with $e_k = 0$ whenever $k < 0$. This dual formulation can be easily deduced from the original version (4.19) through an application of the involution ω. To prove (4.19), we use a combinatorial approach due to [Lindström 73] and [Gessel and Viennot 85]. It is articulated around an interpretation of Schur

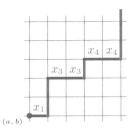

Figure 4.3. The weight of a northeast path.

functions as configurations of nonintersecting northeast lattice paths in $\mathbb{Z} \times \mathbb{Z}$. A *path* $\gamma = (g_1, \ldots, g_n)$ is a sequence of *steps* $(a_{i-1}, b_{i-1}) \xrightarrow{g_i} (a_i, b_i)$ with either

$$(a_i, b_i) = \begin{cases} (a_{i-1}, b_{i-1}) + (1,0), & \text{called an } east \ step, \text{ or} \\ (a_{i-1}, b_{i-1}) + (0,1), & \text{called a } north \ step. \end{cases}$$

Each step of a path is given the *weight*

$$\omega(g_i) = \begin{cases} x_{b_i - b_0 + 1} & \text{if } g_i \text{ is an east step,} \\ 1 & \text{otherwise.} \end{cases} \tag{4.21}$$

The weight of γ is the product of the weights of each of its steps:

$$\mathbf{x}_\gamma := \prod_i \omega(g_i)$$

(see Figure 4.3). We want to consider northeast paths that start at a given (a_0, b_0) and go to $(a_0 + n, \infty)$. Evidently, these paths must end with an infinite sequence of north steps, and their respective weights are monomials of degree n. In fact, there is an obvious bijective correspondence between paths starting at (a, b) and monomials of degree n. We may thus interpret $h_n(\mathbf{x}) = \sum_\gamma \mathbf{x}_\gamma$, as giving the sum of the weighted enumeration of all paths γ going from (a, b) to $(a + n, \infty)$.

In order to get an interpretation for the terms of the determinant

$$\det(h_{\mu_i + j - i})_{1 \le i, j \le n} = \sum_{\sigma \in \mathfrak{S}_n} \operatorname{sgn}(\sigma) h_{\mu_{\sigma(1)} - \sigma(1) + 1} \cdots h_{\mu_{\sigma(k)} - \sigma(k) + k}, \tag{4.22}$$

with $\mu = (\mu_1, \ldots, \mu_k)$ a partition of n, let us consider *k-path configurations* $\Gamma = (\gamma_1, \ldots, \gamma_k)$, with γ_j going from $(-j, 0)$ to $\big(\mu_{\sigma(j)} - \sigma(j), \infty\big)$ for some permutation σ of $\{1, \ldots, k\}$. The weight of such a configuration is defined

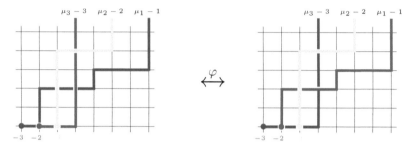

Figure 4.4. Crossing path configuration involution.

to be $\mathbf{x}_\Gamma := \operatorname{sgn}(\sigma) \prod_{i=1}^k \mathbf{x}_{\gamma_i}$, and expanding the right-hand side of (4.22), we get

$$\det\big(h_{\mu_i - i + j}(\mathbf{x})\big)_{1 \le i,j \le n} = \sum_\Gamma \mathbf{x}_\Gamma. \qquad (4.23)$$

The next step is to apply the Lindström–Gessel–Viennot argument to simplify the right-hand side of (4.23). This is accomplished through an involution on the set of k-path configurations considered. This involution is *sign reversing* on the weight of summands that are not fixed by the involution. These correspond to configurations containing some "crossing". Thus, the sum reduces to noncrossing path configurations, since crossing configurations are canceled out by the involution.

To be more precise, we say that $\Gamma = (\gamma_1, \ldots, \gamma_k)$ is *crossing* if there are at least two paths in Γ sharing a common point. We choose s to be the largest integer for which γ_s crosses some other path, and along this path we find the first ($i + j$ minimal) point (i, j) that is a crossing point. We then choose t to be the largest integer for which γ_t crosses γ_s at this point (i, j). In the configuration illustrated in Figure 4.4, the relevant paths are γ_3 and γ_2, respectively starting at $(-3, 0)$ and $(-2, 0)$, and the crossing point considered is $(-2, 0)$. We then construct a new configuration $\varphi(\Gamma) = \big(\varphi(\gamma_1), \ldots, \varphi(\gamma_k)\big)$, first setting $\varphi(\gamma_k) = \gamma_k$, whenever $k \ne s, t$. To describe the effect of φ on the remaining two paths, we consider the respective decompositions of these paths as $\gamma_s = \gamma_s^{(1)} \gamma_s^{(2)}$ and $\varphi(\gamma_t) = \gamma_t^{(1)} \gamma_t^{(2)}$, with $\gamma_s^{(1)}$ and $\gamma_t^{(1)}$ both ending at (i, j). The new configuration is obtained by exchanging end portions of these paths to get two new paths $\varphi(\gamma_s) = \gamma_s^{(1)} \gamma_t^{(2)}$ and $\varphi(\gamma_t) = \gamma_t^{(1)} \gamma_s^{(2)}$. In a more visual format, we have the following picture:

$$
\begin{array}{ccccccccc}
(a_s, b_s) & \to & \cdots & \to & (i, j) & & \cdots & \to & (c_s, d_s) \\
 & & & & & \times & & & \\
(a_t, b_t) & \to & \cdots & \to & (i, j) & & \cdots & \to & (c_t, d_t).
\end{array}
$$

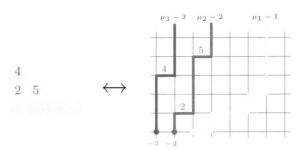

Figure 4.5. Non-crossing configuration associated with a semi-standard tableau.

Exchanging the end portions has the effect of adding ± 1 to the inversion number of the underlying permutation. It follows that $\mathbf{x}_\Gamma = -\mathbf{x}_{\varphi(\Gamma)}$. Starting this process over with the configuration $\varphi(\Gamma)$, we clearly select the same s, (i,j) and t; thus φ is an involution. The whole procedure is illustrated in Figure 4.4. We conclude that

$$\det\big(h_{\mu_i-i+j}(\mathbf{x})\big)_{1\le i,j\le n} = \sum_{\Gamma \text{ noncrossing}} \mathbf{x}_\Gamma.$$

To finish the proof of (4.19), we need only exhibit a weight preserving bijection between noncrossing configurations and semi-standard tableaux of shape μ, thus

$$\sum_{\Gamma \text{ noncrossing}} \mathbf{x}_\Gamma = s_\mu(\mathbf{x}).$$

Given Γ and j ($1 \le j \le k$), consider the list of heights of east steps in γ_{k+1-j}, reading these steps as they occur along the path. The configuration being noncrossing, the length of these lists is increasing as j goes from 1 to k (reading the paths from right to left). We construct the tableau with the jth row given by the jth list obtained. Observe that in a non-crossing configuration the ith east step of γ_j has to occur to the northwest of the ith east step of γ_{j+1}. This implies that the constructed tableau is semi-standard, and this is bijective. The correspondence is illustrated in Figure 4.5.

A similar argument proves the more general version of the Jacobi–Trudi formula for skew Schur functions:

$$s_{\mu/\lambda} = \det(h_{\mu_i-\lambda_j+j-i})_{1\le i,j\le n}, \tag{4.24}$$

or its dual version:

$$s_{\mu'/\lambda'} = \det(e_{\mu_i-\lambda_j+j-i})_{1\le i,j\le n}. \tag{4.25}$$

Examples of direct applications of Jacobi–Trudi formulas are

$$s_{321} = \det \begin{pmatrix} h_3 & h_4 & h_5 \\ h_1 & h_2 & h_3 \\ 0 & 1 & h_1 \end{pmatrix}$$
$$= h_2 h_2 h_3 - h_1^2 h_4 + h_1 h_5 - h_3^2,$$

and

$$s_{4321/21} = \det \begin{pmatrix} h_2 & h_4 & h_6 & h_7 \\ 1 & h_2 & h_4 & h_5 \\ 0 & 1 & h_2 & h_3 \\ 0 & 0 & 1 & h_1 \end{pmatrix}$$
$$= h_1 h_2^3 - 2 h_1 h_2 h_4 - h_2^2 h_3 + h_1 h_6 + h_2 h_5 + h_3 h_4 - h_7.$$

The Jacobi–Trudi identities are useful in many calculations. To illustrate, let us deduce

$$s_\mu[1 - u] = \begin{cases} (-u)^k(1 - u) & \text{if } \mu = (n - k, 1^k), \\ 0 & \text{otherwise,} \end{cases} \tag{4.26}$$

from (3.27). We first expand $s_\mu[1 - u]$ as $\det(h_{\mu_i+j-i}[1 - u])_{1 \le i,j \le n}$, and observe that when $\mu_2 \ge 2$, all entries of the first two rows are equal to $1 - u$, so the determinant vanishes. We are thus reduced to the case of hooks, in which case we have to compute the determinant of $(k + 1) \times (k + 1)$:

$$d_{k+1} := \det \begin{vmatrix} 1 - u & 1 - u & \cdots & 1 - u & 1 - u \\ 1 & 1 - u & \cdots & 1 - u & 1 - u \\ \vdots & \vdots & \ddots & \vdots & \vdots \\ 0 & 0 & \cdots & 1 & 1 - u \end{vmatrix}.$$

Expanding this determinant with respect to the first column, we get the recurrence

$$d_{k+1} = (-u)d_k, \quad \text{with } d_1 = 1 - u,$$

so (4.26) holds.

4.6   Proof of the Hook Length Formula

The objective here is to prove the hook length formula (2.1)

$$f^\mu = \frac{n!}{\prod_{c \in \mu} h(c)}$$

giving the number of standard tableaux of shape μ. Recall that for $c = (i-1, j-1)$ we have $h(c) := (\mu_j - i) + (\mu'_i - j) + 1$. The proof (inspired by one presented in [Manivel 01]) makes use of the number $F_\mu(N)$ of μ-shape semi-standard tableaux with values in the set $\{1, 2, \ldots, N\}$. We begin by linking f^μ to $F_\mu(N)$, through the following argument. From the combinatorial definition (4.1) of Schur functions, we have

$$F_\mu(N) = s_\mu(\underbrace{1, \ldots, 1}_{N \text{ copies}}).$$

Using the Cauchy–Littlewood formula (4.9) with $\mathbf{x} = x_1, x_2, x_3, \ldots$ and $y_j = 1/N$ $(1 \le j \le N)$, we get

$$\sum_\mu s_\mu(\mathbf{x}) \frac{F_\mu(N)}{N^{|\mu|}} = \prod_{i=1}^N \left(1 - \frac{x_i}{N}\right)^{-N},$$

and taking the limit of both sides, as $N \to \infty$, we find that

$$\sum_\mu s_\mu(\mathbf{x}) \lim_{N \to \infty} \frac{F_\mu(N)}{N^{|\mu|}} = \exp\big(h_1(\mathbf{x})\big).$$

However, we already have equation (4.10) giving the Schur expansion h_1^N. This leads to the evaluation

$$\lim_{N \to \infty} \frac{F_\mu(N)}{N^{|\mu|}} = \frac{f^\mu}{|\mu|!}, \tag{4.27}$$

since the $s_\mu(\mathbf{x})$ are linearly independent. The next step consists of computing $F_\mu(N)$ as follows. Setting $x_i = q^{i-1}$ in (3.31), we have

$$s_\mu(1, q, \ldots, q^{n-1}) = \frac{\det\big((q^{\mu_j + n - j})^{i-1}\big)}{\det\big((q^{n-j})^{i-1}\big)}$$

$$= \prod_{i<j} \frac{q^{\mu_j + n - j} - q^{\mu_i + n - i}}{q^{n-j} - q^{n-i}} \tag{4.28}$$

$$= q^{n(\mu)} \prod_{i<j} \frac{1 - q^{\mu_i - \mu_j + j - i}}{1 - q^{j-i}}.$$

This assertion remains true if we add any number of zero parts to μ. We can therefore suppose that N is the number of parts of μ. Observe that the list of hook lengths of cells in the ith row of μ is exactly the complement of the set $\{\mu_i - \mu_j + j - i \mid j > i\}$ relative to the set of numbers between 1 and $\mu_i + N - i$. This is illustrated in Figure 4.6, where we consider

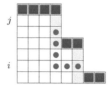

Figure 4.6. Hook shape with endpoints in the rim.

all hook shapes with corner in the ith row and both endpoints in the "rim" of μ. Here, the *rim* of a partition μ is the skew shape μ/ν, where $\nu = (\mu_2 - 1, \ldots, \mu_N - 1)$.

If we also allow hook shapes with top endpoints in position $(0, k + 1)$ up to $(0, N - 1)$, we get exactly one such hook shape for all integers from 1 to $\mu_i + N - i$. Moreover, the length of a hook shape that ends strictly below the top cell in a column is $\mu_i - \mu_j + j - i$, with j standing for the index of the row that sits just above the top end point of the hook shape considered. It follows that

$$\prod_{i<j}(1 - q^{\mu_i-\mu_j+j-i}) = \frac{\prod_{i=1}^{N}\prod_{j=1}^{\mu_i+N-i}(1 - q^j)}{\prod_{c\in\mu}(1 - q^{h(c)})}.$$

We take the quotient of each side of this last identity by the respective sides of

$$\prod_{i<j}(1 - q^{j-i}) = \prod_{i=1}^{N}\prod_{j=1}^{N-i}(1 - q^j),$$

to get

$$s_\mu(1, q, \ldots, q^{N-1}) = q^{n(\mu)} \prod_{(i,j)\in\mu} \frac{1 - q^{N+j-i}}{1 - q^{h(i,j)}}. \tag{4.29}$$

Once again, taking the limit as q tends to 1, we get

$$F_\mu(N) = \prod_{(i,j)\in\mu} \frac{N + j - i}{h(i,j)}.$$

Substituting in (4.27) finally gets us to

$$\frac{f^\mu}{|\mu|!} = \lim_{N\to\infty} \prod_{(i,j)\in\mu} \frac{1 + (j - i)/N}{h(i,j)}$$

$$= \prod_{c\in\mu} \frac{1}{h(c)},$$

showing (2.1). For $|q| < 1$ in (4.29), we can also take the limit as N goes to ∞ to deduce

$$s_\mu(1, q, q^2, \dots) = \frac{q^{n(\mu)}}{\prod_{c \in \mu}(1 - q^{h(c)})}. \tag{4.30}$$

4.7 The Littlewood–Richardson Rule

The *Littlewood–Richardson coefficients*, denoted by $c_{\lambda\mu}^\theta$, are the multiplicative structure constants of the ring $R^\mathfrak{S}$ in the Schur function basis. More precisely, these are the numbers such that

$$s_\lambda s_\mu = \sum_{\theta \vdash |\lambda| + |\mu|} c_{\lambda\mu}^\theta s_\theta. \tag{4.31}$$

They appear in many contexts, such as algebraic geometry, representation theory, theoretical physics, and so on. An a priori surprising fact is that they are nonnegative integers. For example, we have

$$s_{21}s_{32} = s_{53} + s_{521} + s_{44} + 2s_{431} + s_{422} + s_{4211} + s_{332} + s_{3311} + s_{3221}.$$

Below we give a combinatorial recipe for the computation of these integers. Two striking special cases, the *Pieri formulas*, are

$$h_k s_\mu = \sum_\theta s_\theta \quad \text{and} \quad e_k s_\mu = \sum_\theta s_\theta, \tag{4.32}$$

the sum being over the partitions θ for which θ/μ is a k-cell *horizontal strip*, and respectively over those for which θ/μ is a k-cell *vertical strip*. By way of illustration, we have the expansion

with Schur function denoted by the corresponding diagrams. This illustrative device puts the emphasis on the underlying combinatorics.

Another striking feature of the Littlewood–Richardson coefficient is best expressed in terms of adjointness of operators with respect to the scalar product introduced in Section 4.3. For an homogeneous symmetric function f, let us denote by $f^\perp$ the linear operator on symmetric functions that is adjoint to the operation of multiplication by this f. In other words, we have

$$\langle g, fh \rangle = \langle f^\perp g, h \rangle \tag{4.33}$$

for all symmetric functions g and h. The cases $f = h_k$ and $f = e_k$ give rise to the *dual Pieri formulas*

$$h_k^{\perp} s_\mu = \sum_\theta s_\theta \quad \text{and} \quad e_k^{\perp} s_\mu = \sum_\theta s_\theta. \tag{4.34}$$

Here, the sum is over the set of partitions θ that can be obtained from μ by removing a k-horizontal strip (resp. vertical strip). For example, we have

$$h_2^{\perp} s_{4321} = s_{431} + s_{422} + s_{4211} + s_{332} + s_{3311} + s_{3221}.$$

This is also nicely presented in diagram form as

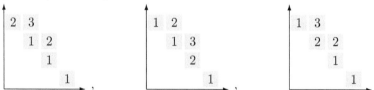

It is a fact that the operator $s_\lambda^{\perp}$ is such that $s_\lambda^{\perp} s_\theta := s_{\theta/\lambda}$, so that the Littlewood–Richardson coefficients actually give the coefficients of the Schur basis expansion of skew Schur functions:

$$s_{\theta/\lambda} = \sum_\mu c_{\lambda\mu}^{\theta} s_\mu. \tag{4.35}$$

Let us now come to the combinatorial description of the coefficients $c_{\lambda\mu}^{\theta}$. The *Littlewood–Richardson rule* states that $c_{\lambda\mu}^{\theta}$ is the number of semi-standard tableaux having skew shape θ/λ and content μ as well as satisfying the further condition that their "reverse reading word is Yamanouchi", in the following sense. The *reverse reading word* $\overleftarrow{w}(\tau)$ of a tableau τ is obtained by reading the entries of τ from left to right along rows, starting with the bottom row. The resulting word is said to be *Yamanouchi* if for all i the number of i is at least as great as the number of $i + 1$ in any initial segment of $\overleftarrow{w}(\tau)$. Let us check that, by this combinatorial rule, 3 is the coefficient of s_{4332} in $s_{321}s_{321}$. Indeed, we have exactly three content 321 Yamanouchi reverse reading word semi-standard tableaux of shape $4332/321$. They are

$$
\begin{array}{cc}
2 & 3 \\
1 & 2 \\
1 & \\
& 1 \\
\end{array}
\qquad
\begin{array}{cc}
1 & 2 \\
1 & 3 \\
2 & \\
& 1 \\
\end{array}
\qquad
\begin{array}{cc}
1 & 3 \\
2 & 2 \\
1 & \\
& 1 \\
\end{array}
$$

All the other reading words of 321-content 4332/321-shape semi-standard tableaux are not Yamanouchi. There are other interesting combinatorial rules (sometimes more efficient) for the computation of Littlewood–Richardson coefficients (see, for instance, [Stanley 97, vol. 2, Appendix 1]).

One of these corresponds to counting the number of shape θ/λ standard tableaux that are jeu de taquin-equivalent to a given standard tableau τ of shape ν.

Exercise. Check the last assertion for some small examples. Describe a natural bijection between the sets involved in the two descriptions.

The following identity is due to [Littlewood 56]:

$$(s_\alpha s_\beta) * s_\theta = \sum_{\lambda, \mu} c^\theta_{\lambda\mu}(s_\alpha * s_\lambda)(s_\beta * s_\mu), \qquad (4.36)$$

with the sum being over the pairs λ, μ for which $c^\theta_{\lambda\mu}$ is not zero, $\lambda \vdash |\alpha|$ and $\mu \vdash |\beta|$. It is worth mentioning, even if it is only to show how to tie together the two different products on symmetric functions. An example of (4.36) is

$$
\begin{aligned}
(s_{21}s_{21}) * s_{42} &= (s_{21} * s_{21})(s_{21} * s_{21}) + (s_{21} * s_{21})(s_{21} * s_3) \\
&\quad + (s_{21} * s_3)(s_{21} * s_{21}) + (s_{21} * s_3)(s_{21} * s_3) \\
&= s_6 + 3s_{51} + 4s_{42} + 5s_{411} + 2s_{33} + 6s_{321} + 5s_{3111} \\
&\quad + 2s_{222} + 4s_{2211} + 3s_{21111} + s_{111111} \\
&\quad + s_{51} + 2s_{42} + 2s_{411} + s_{33} + 4s_{321} + 2s_{3111} + s_{222} + 2s_{2211} + s_{21111} \\
&\quad + s_{51} + 2s_{42} + 2s_{411} + s_{33} + 4s_{321} + 2s_{3111} + s_{222} + 2s_{2211} + s_{21111} \\
&\quad + s_{42} + s_{411} + s_{33} + 2s_{321} + s_{3111} + s_{222} + s_{2211} \\
&= s_6 + 5s_{51} + 9s_{42} + 10s_{411} + 5s_{33} + 16s_{321} + 10s_{3111} \\
&\quad + 5s_{222} + 9s_{2211} + 5s_{21111} + s_{111111}.
\end{aligned}
$$

4.8   Schur-Positivity

Many problems that motivate this work find their origins in the study of "Schur-positivity" of symmetric functions. We have already encountered symmetric functions expressions exhibiting this property: skew-Schur functions, elementary and complete homogeneous symmetric functions, Kronecker product and simple product of symmetric functions. The common feature of all of these is that they have expansions with positive integer coefficients in terms of the Schur function basis. When this is the case, we say that we have *Schur positivity*, or that the function in question is *Schur-positive* (even sometimes just *s-positive*). Illustrating with examples of situations that we have already encountered, we have

$$s_{4321/21} = s_{43} + 2s_{421} + s_{4111} + 2s_{331} + 2s_{322} + 2s_{3211} + s_{2221},$$

$$s_{3111} * s_{321} = s_{51} + 2s_{42} + 2s_{411} + s_{33} + 4s_{321} + 2s_{3111} + s_{222}$$
$$+ 2s_{2211} + s_{21111},$$

$$s_{21}s_{32} = s_{5,3} + s_{5,21} + s_{44} + 2s_{431} + s_{422} + s_{4211} + s_{332}$$
$$+ s_{3311} + s_{3221}.$$

The analogous stronger notions of h-*positive* and e-*positive* are also sometimes considered. As should be expected, these correspond to symmetric
functions whose expansion in the relevant basis has positive integer coefficients. Both cases are seen to be stronger notions than s-positivity by
checking recursively, using (4.32), that h_μ and e_μ are Schur-positive for
all μ.

Checking Schur-positivity is often easier using formal power series with
symmetric function coefficients. We have already encountered such series
in Chapter 3. A typical example is as follows: we want to prove that
the expression $\sum_{k=0}^{\lfloor n/2 \rfloor} p_1^{n-2k} p_2^k$ is Schur-positive for all $n \geq 1$. This is not
immediate since $p_2 = s_2 - s_{11}$. The expressions under consideration clearly
appear as the coefficients of ζ^n in the series

$$\frac{1}{(1 - p_1\zeta)(1 - p_2\zeta^2)} = 1 + p_1\zeta + (p_1^2 + p_2)\zeta^2 + (p_1^3 + p_1p_2)\zeta^3$$
$$+ (p_1^4 + p_2p_1^2 + p_2^2)\zeta^4 + \cdots.$$

We verify that this series is indeed Schur-positive,[3] as follows. First, we
calculate the series identity

$$\frac{1}{(1 - p_1\zeta)(1 - p_2\zeta^2)} = \frac{1}{(1 - s_1\zeta)\big(1 - (s_2 - s_{11})\zeta^2\big)}$$
$$= \frac{1 + s_1\zeta}{(1 - s_1^2\zeta^2)\big(1 - (s_2 - s_{11})\zeta^2\big)}$$
$$= \frac{1 + s_1\zeta}{\big(1 - (s_2 + s_{11})\zeta^2\big)\big(1 - (s_2 - s_{11})\zeta^2\big)}$$
$$= \frac{1 + s_1\zeta}{2s_{11}\zeta^2}\left(\frac{1}{1 - (s_2 + s_{11})\zeta^2} - \frac{1}{1 - (s_2 - s_{11})\zeta^2}\right).$$

Then we observe that the large expression in parentheses in the final step of
this calculation is simply the sum of the terms that contain odd powers of
s_{11} in the manifestly Schur-positive series $\big(1 - (s_2 + s_{11})\zeta^2\big)^{-1}$. We conclude
that we have expressed $\sum_{k=0}^{\lfloor n/2 \rfloor} p_1^{n-2k} p_2^k$ as a positive integer polynomial in
s_1, s_2, and s_{11}.

---

[3]Naturally extending the notion of Schur-positivity to such series.

Exercise. Show that the coefficient of ζ^n is Schur positive in

$$\frac{1}{(1 - p_1\,\zeta)\,(1 - p_2\,\zeta^2)\,(1 - p_3\,\zeta^3)}.$$

Other tools for the proof of Schur-positivity will be introduced in the upcoming chapters. It will become apparent, as we go along, that we need to consider a broader notion of Schur-positivity. Symmetric functions over the field of rational fraction in two formal parameters q and t will become the norm. In that context we will say that we have Schur-positivity if the relevant coefficients of Schur functions lie in $\mathbb{N}[q, t]$. This is to say that they are positive integer polynomials in q and t.

4.9   Poset Partitions

For our upcoming discussion of quasisymmetric functions we need to recall some notions and notations regarding π-partitions for some n-*poset*[4] π. We write $a \prec b$ if a is less than b with respect to the poset order. A poset is typically describe by its *covering relation* or *Hasse diagram*. Recall that this is the oriented graph with an arc going from a to b if and only if a is *covered* by b in π. This is to say that $a \prec b$ and that if there is some c such that $a \preceq c \preceq b$ then it can only be because either $c = a$ or $c = b$. Hasse diagrams are usually drawn with arcs pointing upward so that arc orientation need not be specified (e.g., Figure 4.7).

For our presentation we also need an integer labeling for the elements of the n-poset π. This comes from a given bijection between the underlying set of π and the set $\{1, \ldots, n\}$. In the next definition we identify elements of π with their label. Let us say that $f : \pi \longrightarrow \mathbb{N}^+$ is a π-*partition* of the integer N if

(1) $f(a) \geq f(b)$, whenever $a \preceq_\pi b$ and $a < b$ (as integers),

(2) $f(a) > f(b)$, whenever $a \preceq_\pi b$ and $a > b$ (as integers),

(3) $f(1) + \cdots + f(n) = N$.

The values $f(i)$ are the *parts* of the π-partition f. Observe that we get back the usual notion of length k partition in the special case of π being the set $\{1, 2, \ldots, k\}$ with its usual order.

We will be mostly interested in posets constructed from n-cell diagrams

$$\mathbf{d} = \{(a_1, b_1), (a_2, b_2), \ldots, (a_n, b_n)\}$$

---

[4] A partially ordered set of order n.

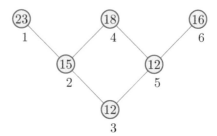

Figure 4.7. A poset partition for $\pi = 321$.

as follows. We set $(a, b) \preceq (c, d)$ exactly when $a \leq c$, and $b \leq d$. We successively affix a label between 1 and n to the cells of $\mathbf{d}$ proceeding column by column, from left to right, and going from top to bottom within each column. The resulting poset is also denoted by $\mathbf{d}$. Observe that these conventions make it so that (in a $\mathbf{d}$-partition) part sizes increase strictly along columns and weakly along rows. Thus, they correspond to $\mathbf{d}$-shape semi-standard tableaux in the sense of Section 2.1. In particular, poset partitions for a skew-shape λ/μ are exactly the semi-standard tableaux of shape λ/μ. To illustrate, we may consider the poset partition associated with the Young diagram of 321 as depicted in Figure 4.7. Here, labels appear just under the vertices and corresponding part sizes are the numbers appearing inside them. We denote by $\mathcal{A}(\pi)$ the set of poset partitions on the poset π. This set can be decomposed in a natural fashion using the following two notions. For a given permutation $\sigma \in \mathfrak{S}_n$, we say that a function $f \colon \{1, \dots, n\} \to \mathbb{N}$ is σ-*compatible* if

$$f\big(\sigma(1)\big) \leq f\big(\sigma(2)\big) \leq \cdots \leq f\big(\sigma(n)\big)$$

and $f\big(\sigma(i)\big) < f\big(\sigma(i+1)\big)$ whenever $\sigma(i) > \sigma(i+1)$. The set $\mathcal{L}(\pi)$ of *linear extensions* of an n-element poset π is the set of permutations σ such that $\sigma(i) < \sigma(j)$ if $i \prec j$. It is straightforward to check that

$$\mathcal{A}(\pi) = \sum_{\sigma \in \mathcal{L}(\pi)} \mathcal{C}(\sigma), \tag{4.37}$$

with $\sum$ denoting disjoint union and $\mathcal{C}(\sigma)$ standing for the set of σ-compatible functions.

4.10   Enumeration of Poset Partitions and Quasisymmetric Functions

Our next step is to consider a weighted enumeration of poset partitions. For an infinite (denumerable) set of commuting variables $\mathbf{x} = x_1, x_2, \dots,$

we consider the formal series $s_\pi(\mathbf{x}) := \sum_{f \in \mathcal{A}(\pi)} \mathbf{x}_f$, where $\mathbf{x}_f$ denotes the monomial $x_{f(1)} x_{f(2)} \cdots x_{f(n)}$. In view of the remark at the end of the last section, this gives a natural generalization of Schur functions. However, for diagrams that do not correspond to skew partitions, the function $s_\pi(\mathbf{x})$ need not be a symmetric. It is rather a quasisymmetric function in the following sense.

We begin by defining the *monomial quasisymmetric functions* (or *polynomials* if $\mathbf{x} = x_1, \ldots, x_n$) as

$$M_\mathbf{a} = M_\mathbf{a}(\mathbf{x}) := \sum_{\substack{\mathbf{y} \subseteq \mathbf{x} \\ \#\mathbf{y} = \ell(\mathbf{a})}} \mathbf{y}^\mathbf{a}, \tag{4.38}$$

with $\mathbf{y}$ running over all $|\mathbf{a}|$-subsets of $\mathbf{x}$. These functions are indexed by compositions, and we suppose that the order on the $\mathbf{y}$ variables is induced from that on the $\mathbf{x}$ variables. Thus, the monomial $\mathbf{y}^\mathbf{a}$ is well defined. To illustrate definition (4.38), we have

$$M_{21}(\mathbf{x}) = x_1^2 x_2 + x_1^2 x_3 + \cdots + x_a^2 x_b + \cdots,$$

where $a < b$ in the general term. Observe that any symmetric function can be expanded in terms of the linear basis of monomial quasisymmetric functions, since by definition we have $m_\mu = \sum_{\hat{\mathbf{a}} = \mu} M_\mathbf{a}$. We define *quasisymmetric functions* to be any finite linear combinations of the $M_\mathbf{a}$. It turns out that they constitute a subalgebra of $R = \mathbb{C}[\mathbf{x}]$, which is denoted by $R^{\sim \mathfrak{S}_n}$ when we have n variables, or $R^{\sim \mathfrak{S}}$ if n goes to infinity. The fact that this space is closed under multiplication is not directly obvious from the definition. This will be discussed in Section 4.11.

Observe that $\mathbf{x}^\mathbf{a}$ is the leading monomial of $M_\mathbf{a}(\mathbf{x})$ in the lexicographic order for monomials. This implies that the nonzero[5] $M_\mathbf{a}(\mathbf{x})$ are linearly independent. It may also be worth observing that we have the expansion formula

$$f(\mathbf{x}) = \sum_\mathbf{a} (f, \mathbf{x}^\mathbf{a}) M_\mathbf{a}(\mathbf{x}), \tag{4.39}$$

with $(f, \mathbf{x}^\mathbf{a})$ standing for the coefficient of the monomial $\mathbf{x}^\mathbf{a}$ in f. It follows that

$$\mathrm{Hilb}_q(R^{\sim \mathfrak{S}_n}) = 1 + \frac{q}{1-q} + \cdots + \frac{q^n}{(1-q)^n}, \tag{4.40}$$

since $q^k/(1-q)^k$ is the generating series for length k compositions.

---

[5] We must be careful when $\mathbf{a}$ has more parts than the number of variables in $\mathbf{x}$.

Fundamental Basis

Another basis of the space of quasisymmetric functions (or polynomials) plays a crucial role in our story: the *fundamental basis* $Q_\mathbf{a}$. For a composition $\mathbf{a}$, it is defined by setting

$$Q_\mathbf{a}(\mathbf{x}) := \sum_{\mathbf{b} \preceq \mathbf{a}} M_\mathbf{b},$$

writing $\mathbf{b} \preceq \mathbf{a}$ whenever $\mathbf{b}$ is a refinement of $\mathbf{a}$ (see Section 1.8). The reverse relation, expressing the $M_\mathbf{a}$ in terms of the $Q_\mathbf{a}$, is easily obtained by a Möbius inversion argument giving

$$M_\mathbf{a}(\mathbf{x}) := \sum_{S_\mathbf{a} \subseteq T \subseteq \{1,\dots,n-1\}} (1-)^{\#T - \#S_\mathbf{a}} Q_{\mathrm{co}(T)}(\mathbf{x}). \tag{4.41}$$

Once again, see Section 1.8 for the notations $S_\mathbf{a}$ and $\mathrm{co}(T)$. As we did for symmetric functions, we often omit mention of the variables, essentially assuming that they are in infinite number. For example, we have

$$Q_{32} = M_{32} + M_{311} + M_{212} + M_{2111} + M_{122} + M_{1211} + M_{1112} + M_{11111},$$
$$M_{32} = Q_{32} - Q_{311} - Q_{212} + Q_{2111} - Q_{122} + Q_{1211} + Q_{1112} - Q_{11111},$$

although some of the terms may vanish if we have too few variables. One of the manifold uses of the fundamental basis is the following immediate consequence of the definitions

$$\sum_{f \in \mathcal{C}(\sigma)} \mathbf{x}_f = Q_{\mathrm{co}(\mathrm{Des}(\sigma))}, \tag{4.42}$$

with f running over the set of σ-compatible functions $f \colon \{1,\dots,n\} \to \mathbb{N}$, and $\mathrm{Des}(\sigma)$ denoting the descent set of the permutation σ as defined in Section 1.1.

Main Enumeration Result

We can now obtain the weighted enumeration mentioned at the start of this section. Indeed, we have the following very nice result, which can be found in [Stanley 97, (7.94)].

Proposition 4.1. *For all labelled posets π, the poset partition enumerator $s_\pi(\mathbf{x})$ is a quasisymmetric series. In fact, we have the expansion*

$$s_\pi(\mathbf{x}) = \sum_{\sigma \in \mathcal{L}(\pi)} Q_{\mathrm{co}(\sigma)}(\mathbf{x}), \tag{4.43}$$

using the convention of writing $\mathrm{co}(\sigma)$ for $\mathrm{co}\big(\mathrm{Des}(\sigma)\big)$.

In the special case of a poset associated with a skew partition, we can reformulate this proposition to get an expansion of any skew Schur function in terms of the fundamental basis:

$$s_{\lambda/\mu} = \sum_{\tau} Q_{\mathrm{co}(\tau)}. \tag{4.44}$$

Here, τ varies in the set of all standard tableaux of shape λ/μ, and $\mathrm{co}(\tau)$ is the composition associated with the reading descent set of τ. Recall that i is said to be a reading descent of τ if $i+1$ sits to the northwest of i in τ. For example, 325 is the composition associated with the tableau

$$\begin{array}{llll} 6 & 7 & & \\ 1 & & & \\ & 4 & 5 & 8 \\ & & 2 & 3 & 9 & 10 \end{array}$$

since its reading descents correspond to the entries 3 and 5. From this formula, we can deduce (see [Stanley 97, Proposition 7.19.11]) that

$$s_{\lambda/\mu}[1/(1-q)] = \frac{\sum_{\tau} q^{\mathrm{rmaj}(\tau)}}{(1-q)(1-q^2)\cdots(1-q^n)}, \tag{4.45}$$

with τ varying in the set of standard tableau of shape λ/μ. Here, the *reading major index*, $\mathrm{rmaj}(\tau)$, is just the sum of all the i in τ that are reading descents.

More on the General Diagram $s_{\mathbf{d}}(\mathbf{x})$

Even more generally, let us consider the analogous weighted enumeration of semi-standard $\mathbf{d}$-shape tableaux, resulting in the quasisymmetric function $s_{\mathbf{d}}$. For instance, the only possible semi-standard fillings of the diagram $\mathbf{d} = (0,0), (1,0), (1,1)$ are as shown in Figure 4.8 for integers $a < b < c$. Summing the corresponding monomials for all possible choices of a, b and c, we see that $s_{\mathbf{d}} = Q_{21}$. This example is somewhat particular in that it is a special case of the notion of a *forward ribbon*, which are n-cell diagrams $r(\mathbf{a})$ recursively associated with n-compositions as follows. We begin by

$$\begin{array}{cccc} b & & c & \\ a & a & \quad a & b \end{array}$$

Figure 4.8. Semi-standard d-shape tableaux.

Figure 4.9. Forward ribbon associated with $(3, 1, 4, 2)$.

setting $r(0) = \emptyset$. Now, let c be the last part of a length $k + 1$ composition **a**, and let **b** be the composition obtained from **a** by removing this last part. Then the diagram $r(\mathbf{a})$ is (recursively) obtained by adding to the diagram of $r(\mathbf{b})$ the cells

$$(m, k), (m + 1, k), \ldots, (m + c - 1, k),$$

with m equal to $n - c - k + 1$. Illustrating this process with the composition $\mathbf{a} = (3, 1, 4, 2)$, we get the forward ribbon of Figure 4.9.

Exercise. Show that, for all compositions **a**, we have $s_{r(\mathbf{a})} = Q_{\mathbf{a}}$.

4.11   Multiplicative Structure Constants

Our description of the multiplication rule for the basis of monomial quasisymmetric functions is in terms of the *quasi-shuffle* bilinear map

$$\tilde{\sqcup} : \mathbb{Q}\mathcal{C} \times \mathbb{Q}\mathcal{C} \longrightarrow \mathbb{Q}\mathcal{C}$$

(see [Hoffman 00]) on the free $\mathbb{Q}$-vector space generated by the set $\mathcal{C}$ of compositions, here written as words. This operation is described recursively using an extension of concatenation $\mathbf{a} \cdot \mathbf{b}$ to a bilinear operation on $\mathbb{Q}\mathcal{C}$. For the empty composition 0, we begin by setting $\mathbf{c} \tilde{\sqcup} 0 := \mathbf{c}$ and $0 \tilde{\sqcup} \mathbf{c} := \mathbf{c}$. For integers a and b and compositions **c** and **d**, the general recursive statement is that

$$(a \cdot \mathbf{c}) \tilde{\sqcup} (b \cdot \mathbf{d}) := a \cdot (\mathbf{c} \tilde{\sqcup} b \cdot \mathbf{d}) + b \cdot (a \cdot \mathbf{c} \tilde{\sqcup} \mathbf{d}) + e \cdot (\mathbf{c} \tilde{\sqcup} \mathbf{d}), \qquad (4.46)$$

writing e for the integer $a + b$ (to make it clear that the right-hand side is a formal linear combination of compositions). For instance, we have

$$12 \tilde{\sqcup} 32 = 1 \cdot (2 \tilde{\sqcup} 32) + 3 \cdot (12 \tilde{\sqcup} 2) + 4 \cdot (2 \tilde{\sqcup} 2)$$
$$= 1232 + 1322 + 1322 + 134 + 152$$
$$+ 3122 + 3122 + 3212 + 332 + 314$$
$$+ 422 + 422 + 44.$$

With this notation at hand, the multiplication rule (see [Stanley 79]) for the monomial basis is simply

$$M_{\mathbf{a}}M_{\mathbf{b}} = \sum_{\mathbf{c}} \langle \mathbf{c} \mid \mathbf{a} \mathbin{\tilde{\sqcup}} \mathbf{b} \rangle M_{\mathbf{c}}, \qquad (4.47)$$

with $\langle \mathbf{c} \mid \mathbf{a} \mathbin{\tilde{\sqcup}} \mathbf{b} \rangle$ corresponding to the coefficient of $\mathbf{c}$ in $\mathbf{a} \mathbin{\tilde{\sqcup}} \mathbf{b}$. In view of our calculation above, we get

$$M_{12}M_{32} = M_{1232} + 2M_{1322} + M_{134} + M_{152} + 2M_{3122} + M_{3212}$$
$$+ M_{332} + M_{314} + 2M_{422} + M_{44}.$$

An alternate description of the ring structure of $R^{\sim \mathfrak{S}_n}$ can be given in terms of the basis $\{P_{\mathbf{a}}\}_{\mathbf{a}}$, introduced in [Malvenuto and Reutenauer 95]. In this case, the multiplication rule is directly expressed in terms of the usual notion of shuffle. To describe this new basis, we need only specify that the relevant changes of bases are given by the formulas

$$P_{\mathbf{a}} = \sum_{\mathbf{a} \geq \mathbf{b}} \frac{1}{f(\mathbf{a}, \mathbf{b})} M_{\mathbf{b}} \quad \text{and} \quad M_{\mathbf{a}} = \sum_{\mathbf{a} \geq \mathbf{b}} \frac{1}{g(\mathbf{a}, \mathbf{b})} P_{\mathbf{b}}, \qquad (4.48)$$

with $f(\mathbf{a}, \mathbf{b})$ and $g(\mathbf{a}, \mathbf{b})$ integers calculated as follows. Let us suppose that

($*$) the given composition $\mathbf{a}$ can be written as a concatenation $\mathbf{a}^{(1)} \cdots \mathbf{a}^{(k)}$ of k compositions $\mathbf{a}^{(i)}$ such that we have $\mathbf{a}^{(i)} \models b_i$, with $\mathbf{b} = b_1 \cdots b_k$.

When this is the case, we set

$$f(\mathbf{a}, \mathbf{b}) := \ell(\mathbf{a}^{(1)})! \cdots \ell(\mathbf{a}^{(k)})!,$$

and

$$g(\mathbf{a}, \mathbf{b}) := (-1)^{\ell(\mathbf{a})-k} \ell(\mathbf{a}^{(1)}) \cdots \ell(\mathbf{a}^{(k)}).$$

In equation (4.48), the sums are over the set of compositions $\mathbf{b}$ for which condition ($*$) makes sense. For example, considering the compositions of 3, we get

$$P_3 = M_3,$$

$$P_{21} = M_{21} + \frac{1}{2}M_3,$$

$$P_{12} = M_{12} + \frac{1}{2}M_3, \qquad (4.49)$$

$$P_{111} = M_{111} + \frac{1}{2}M_{21} + \frac{1}{2}M_{12} + \frac{1}{6}M_3.$$

For the P-basis, the multiplication rule is simply expressed as

$$P_{\mathbf{a}}P_{\mathbf{b}} = \sum_{\mathbf{c}} \langle \mathbf{c} \mid \mathbf{a} \sqcup \mathbf{b} \rangle P_{\mathbf{c}}. \qquad (4.50)$$

Exercise. Show the product rule

$$Q_{\mathbf{a}}Q_{\mathbf{b}} = Q_{\mathbf{ab}} + Q_{\mathbf{a}\oplus\mathbf{b}},$$

where

$$(a_1,\ldots,a_{k-1},a_k) \oplus (b_1,b_2,\ldots,b_\ell) := (a_1,\ldots,a_{k-1},a_k+b_1,b_2,\ldots,b_\ell).$$

Using this multiplication rule, expand Q_1^n in terms of the $Q_{\mathbf{a}}$.

4.12 r-Quasisymmetric Polynomials

An interesting variant of the notion of quasisymmetric polynomial, due to [Hivert 08], consists of considering elements of the vector space freely spanned by the set of *monomial r-quasisymmetric polynomials* defined as

$$M_{\mathbf{a},\lambda}(\mathbf{x}) := \sum_{\substack{\mathbf{y}+\mathbf{z}=\mathbf{x} \\ \#\mathbf{y}=\ell(\mathbf{a})}} \mathbf{y}^{\mathbf{a}} m_\lambda(\mathbf{z}). \tag{4.51}$$

Here, r is some integer ≥ 1, and $\mathbf{y} + \mathbf{z} = \mathbf{x}$ means $\mathbf{y}$ is some subset of the set of variables, with $\mathbf{z}$ its complement in $\mathbf{x}$. Monomial r-quasisymmetric polynomials are indexed by pairs $(\mathbf{a}, \lambda)$, where $\mathbf{a}$ is an "r-composition" and λ is a partition with part sizes $< r$. An *r-composition* is a composition with all parts $\geq r$. In a compact format this property on parts of $\mathbf{a}$ will be written $\mathbf{a} \geq r$. An example of a monomial r-quasisymmetric polynomial, with $r = 2$ and $n = 4$, is

$$M_{23,1} = x_1^2 x_2^3 (x_3 + x_4) + x_1^2 x_3^3 (x_2 + x_4) + x_1^2 x_4^3 (x_2 + x_3)$$
$$+ x_2^2 x_3^3 (x_1 + x_4) + x_2^2 x_4^3 (x_1 + x_3) + x_3^2 x_4^3 (x_1 + x_2).$$

We write $M_{\mathbf{a}}$ when λ is the empty partition, and we say that in that case $M_{\mathbf{a}}$ is *partition free*. An *r-quasisymmetric polynomial* is a polynomial that can be written as a linear combination of monomial r-quasisymmetric polynomials. Clearly the notion of 1-quasisymmetric polynomial reduces to the usual notion of quasisymmetric polynomial.

Assuming that the space $R^{\sim_r \mathfrak{S}_n}$ of r-quasisymmetric polynomials is closed under multiplication (which is indeed the case, see [Hivert 08]), it is clear that $R^{\mathfrak{S}_n}$ is a subring of $R^{\sim_r \mathfrak{S}_n}$.

Exercise. Describe a multiplication rule for r-quasisymmetric polynomials written in the M-basis. Also, describe a multiplication rule for the basis obtained by replacing the $m_\lambda(\mathbf{z})$ by the Schur functions $s_\lambda(\mathbf{z})$ in Definition (4.51).

Chapter 5

Some Representation Theory

Our intention in this chapter is to briefly recall the essential notions and results of representation theory of finite groups, especially in the case of the symmetric group. Following Frobenius, we bijectively associate a symmetric function to characters of representations of $\mathfrak{S}_n$. It turns out that, under this natural passage to symmetric functions, Schur functions correspond to irreducible representations. It follows that symmetric functions associated with representations are Schur-positive, with coefficients corresponding to multiplicities of irreducible representations. This will be a recurring theme throughout the rest of this book. For proofs of the results outlined here, we refer to [Sagan 91], or the now classic [Fulton and Harris 91]. Many other notions not presented here are found in [Goodman and Wallach 98].

5.1   Basic Representation Theory

For our general presentation of representation theory it is much simpler to assume that we are working over an algebraic closed field of characteristic zero. In the symmetric group case however, everything works fine over $\mathbb{Q}$, and most of the representations of $\mathfrak{S}_n$ considered in the sequel appear as finite-dimensional sub $\mathfrak{S}_n$-*modules* of of the ring of polynomials $\mathbb{Q}[\mathbf{x}]$. These submodules are all homogeneous, and this plays an important role in our discussion.

For a finite-dimensional vector space $\mathcal{V}$, the *general linear group* $\mathrm{GL}(\mathcal{V})$ is the group of linear isomorphisms of $\mathcal{V}$. A *linear representation* of a finite group G is simply a group homomorphism $\rho\colon G \longrightarrow \mathrm{GL}(\mathcal{V})$. In equivalent terms, this defines a *linear action* $\rho\colon G \times \mathcal{V} \longrightarrow \mathcal{V}$ of G on $\mathcal{V}$, usually denoted as a left multiplication $(g, v) \mapsto g \cdot v$, and defined as satisfying the following:

(1) $1 \cdot v = v$,

(2) $g \cdot (av + bw) = ag \cdot v + bg \cdot w$,

(3) $g_1 \cdot (g_2 \cdot v) = (g_1 g_2) \cdot v$.

This leads to the terminology: $\mathcal{V}$ is a G-*module*. The *dimension* of the representation is the dimension of the vector space $\mathcal{V}$. A *homomorphism* of G-modules is a linear transformation $\theta \colon \mathcal{V} \longrightarrow \mathcal{W}$ that is compatible with the respective G-actions on $\mathcal{V}$ and $\mathcal{W}$, i.e., $\theta(g \cdot v) = g \cdot \theta(v)$. Naturally, we say that θ is an *isomorphism* when it is bijective. A G-*invariant subspace* of a G-module $\mathcal{V}$, is a subspace $\mathcal{U}$ such that $G \cdot \mathcal{U} \subseteq \mathcal{U}$. The *direct sum* of two G-modules $\mathcal{V}$ and $\mathcal{W}$ is equipped with the componentwise G-action. A G-module is said to be *irreducible* if it is not isomorphic to a sum of two (nontrivial) G-modules. Equivalently, it is irreducible if and only if it has no nontrivial G-invariant submodule.

The two basic results of representation theory are as follows. The first states that there is a finite number of nonisomorphic irreducible representations of G. This number is the number of conjugacy classes of G. Let us denote by $\mathcal{C}$ the set of these conjugacy classes. A *complete system* of representatives of irreducible G-modules is of the form $\{\mathcal{V}^c \mid c \in \mathcal{C}\}$, with each $\mathcal{V}^c$ irreducible and $\mathcal{V}^c \not\simeq \mathcal{V}^d$ if $c \neq d$. In the case of the symmetric group $\mathfrak{S}_n$ conjugacy classes are naturally indexed by partitions of n.

The second basic result states that every G-module $\mathcal{W}$ decomposes "uniquely" into irreducible representations: $\mathcal{W} \simeq \sum_{c \in \mathcal{C}} a_c \mathcal{V}^c$. In this decomposition of $\mathcal{W}$ as a direct sum of irreducible G-modules, several copies of a given irreducible G-module $\mathcal{V}_c$ may appear. The maximal number of linearly independent copies of $\mathcal{V}^c$ in $\mathcal{W}$ is the *multiplicity* a_c of this irreducible component in the G-module considered. The problem of finding the irreducible decomposition of $\mathcal{W}$ is thus turned into the "combinatorial" problem of computing these multiplicities. This may be very hard to do in some instances. Observe that we have

$$\dim(\mathcal{W}) = \sum_{c \in \mathcal{C}} a_c \dim(\mathcal{V}^c). \tag{5.1}$$

5.2   Characters

One striking aspect of representation theory is that all the necessary information concerning a G-module is encoded in its *character* $\chi = \chi_{\mathcal{V}}$. This is the function $\chi \colon G \longrightarrow \mathbb{C}$ defined for g in G as the trace $\chi_{\mathcal{V}}(g) := \mathrm{Trace}(g)$ of the linear transformation $g \colon \mathcal{V} \longrightarrow \mathcal{V}$. The importance of characters lies in the fact that two G-modules are isomorphic if and only if they have the same characters. In other words, a G-module is "characterized" by its character.[1] Observe that the value of the character at the identity of G is the dimension of $\mathcal{V}$.

Characters of representations are easily seen to be constant on conjugacy classes of G, since $\text{Trace}(h^{-1}gh) = \text{Trace}(g)$. The characters of irreducible representations are said to be *irreducible characters*. The functions on $\mathfrak{S}_n$ that are constant on conjugacy classes form the *space of central functions* $\mathcal{C}(\mathfrak{S}_n)$. The irreducible characters form a basis of $\mathcal{C}(\mathfrak{S}_n)$. If, on $\mathcal{C}(\mathfrak{S}_n)$ we consider the scalar product

$$\langle \chi, \xi \rangle = \frac{1}{|G|} \sum_{g \in G} \chi(g)\overline{\xi(g)},$$

then the irreducible characters form an orthonormal basis. A further fundamental property of characters is that $\chi_{\mathcal{V} \oplus \mathcal{W}} = \chi_{\mathcal{V}} + \chi_{\mathcal{W}}$. It follows that $\chi_{\mathcal{V}} = \sum_{c \in \mathcal{C}} a_c \chi^c$, with $a_c = \langle \chi_{\mathcal{V}}, \chi^c \rangle$. In particular, a character χ is irreducible if and only if $\langle \chi, \chi \rangle = 1$.

5.3 Special Representations

The simplest representation of a group G is certainly the *trivial representation* $\mathbf{1}$, which sends all elements of G to the identity map of a one-dimensional vector space $\mathcal{V}$. Its character is clearly such that $\chi_1(g) = 1$, for all g in G. It is clearly an irreducible representation, as are all one-dimensional representations.

In Section 1.1 we (unknowingly) encountered two representations of $\mathfrak{S}_n$. These were the *defining representation* that associates the corresponding permutation matrix M_σ with σ, and the *sign representation* ε whose character has value $\chi_\varepsilon(\sigma) = \varepsilon(\sigma)$. The value at σ of the character of the defining representation is the number of fixed points of σ.

A more intricate representation is the (left) *regular representation* $\mathcal{R}$. This has as underlying vector space the free vector space $\mathbb{C}[G]$ spanned by G, and it is turned into a G-module by left-multiplication:

$$h \cdot \sum_{g \in G} a_g g := \sum_{g \in G} a_g hg.$$

---

[1] It may appear strange that we get all this information out of so little knowledge. However, for any $k \geq 0$, the character contains the information about the value of $\text{Trace}(g^k) = \lambda_1^k + \cdots + \lambda_n^k$, with the λ_i being the eigenvalues of g. Thus, if g is diagonalizable, we can reconstruct g up to matrix conjugation, since its eigenvalues are the roots of the characteristic polynomial of g, whose coefficients can be calculated from the knowledge of $\text{Trace}(g^k)$.

The dimension of $\mathcal{R}$ is $|G|$, and its character is

$$\chi_{\mathcal{R}}(g) = \begin{cases} |G| & \text{if } g = \text{Id}, \\ 0 & \text{otherwise.} \end{cases} \tag{5.2}$$

This is easily checked by considering the matrix of left multiplication by g in the basis of $\mathcal{R}$ corresponding to elements of g. Indeed, the only nonzero contribution to $\text{Trace}(g)$ comes from elements g such that $gh = h$ for some h, forcing $g = \text{Id}$. One other nice general result of representation theory is that all irreducible representations of G appear in $\mathcal{R}$ with multiplicity equal to their dimension. In symbols,

$$\mathcal{R} \simeq \bigoplus_{c \in \mathcal{C}} d_c \mathcal{V}^c, \tag{5.3}$$

where the direct sum is over the set $\mathcal{C}$ of conjugacy classes of G, and d_c stands for the dimension of $\mathcal{V}^c$. Taking the dimension of both sides of (5.3), we get the nice identity

$$|G| = \sum_{c \in \mathcal{C}} d_c^2. \tag{5.4}$$

We will soon make all this more explicit in the case of the symmetric group.

Another interesting natural action of a group G comes from the linear extension of conjugation, $g \cdot h := g^{-1}hg$, to $\mathbb{C}[G]$. Evidently the resulting *conjugating representation* has the same dimension as that of the regular representation, since both share the same underlying space $\mathbb{C}[G]$.

Exercise. Show that the conjugating representation of a group G contains at least as many copies of the trivial representation as there are conjugacy classes in G.

5.4   Action of $\mathfrak{S}_n$ on Bijective Tableaux

We now consider an action of $\mathfrak{S}_n$ on the set of bijective fillings of an n-cell diagram $\mathbf{d}$ by elements of the set $\{1, 2, \ldots, n\}$. These are called *bijective tableaux* of shape $\mathbf{d}$. We act on a bijective tableau $\tau \colon \mathbf{d} \longrightarrow \{1, 2, \ldots, n\}$ by left composition $\sigma \cdot \tau$ of permutations σ in $\mathfrak{S}_n$. Thus, for all $c \in \mathbf{d}$ we set $(\sigma \cdot \tau)(c) := \sigma\big(\tau(c)\big)$. This action is illustrated in Figure 5.1. We denote by Col_τ the *column-fixing group* of τ. Its elements are the permutations that fix all the rows of τ, i.e., $\tau(c) = (\sigma \cdot \tau)(c')$ implies that c and c' lie in the same column of $\mathbf{d}$. The conjugates of this subgroup are easily seen to satisfy the identity $\text{Col}_{\sigma \cdot t} = \sigma \, \text{Col}_\tau \, \sigma^{-1}$. In a similar way, we have Row_τ, the *row-fixing group* of τ. To each $\mathbf{d}$-shape bijective tableau τ, let

Figure 5.1. Action of $\sigma = 7152436$ on a bijective tableau.

$$
\begin{array}{cccc}
7 & & & \\
5 & 6 & & \\
1 & 2 & 3 & 4
\end{array}
$$

Figure 5.2. Row reading tableau of $\mu = 421$.

us associate[2] the *tableau monomial* $\mathbf{x}^\tau := \prod_{(a,b)\in\mathbf{d}} x^b_{\tau(a,b)}$. Evidently, this monomial characterizes τ up to a row-fixing permutation of its entries, since the exponent of a variable x_i encodes the row on which the value i lies. For the first tableau of Figure 5.1 we have $\mathbf{x}^\tau = x_2 x_6 x_1^2 x_5^2$, whereas for the second we get $\mathbf{x}^\tau = x_1 x_3 x_4^2 x_7^2$. The action of $\mathfrak{S}_n$ on bijective tableaux is compatible with the action on monomials, i.e., $\sigma \cdot \mathbf{x}^\tau = \mathbf{x}^{\sigma \cdot \tau}$, and row-fixing permutations of τ fix the corresponding monomial $\mathbf{x}^\tau$.

We specialize these considerations to n-cell partition diagrams μ and consider the $\mathfrak{S}_n$-modules

$$
\mathcal{H}^\mu := \mathbb{C}[\mathbf{x}^\tau \mid \tau \text{ bijective tableau of shape } \mu].
$$

This notion is essentially a direct translation to the context of polynomials of the notion of "tabloids" in [Sagan 91]. Let us choose one fixed bijective tableau τ of shape μ. Using this fixed tableau, we consider the module $\mathbb{C}\mathfrak{S}_n\mathbf{x}^\tau := \mathbb{C}[\sigma \cdot \mathbf{x}^\tau \mid \sigma \in \mathfrak{S}_n]$. Clearly, we have the equality $\mathcal{H}^\mu = \mathbb{C}\mathfrak{S}_n\mathbf{x}^\tau$, since $\mathfrak{S}_n$ acts transitively on bijective tableaux of shape μ. It is typical to pick τ to be the *row reading tableau* of shape μ. This is the tableau obtained by filling (i,j) with the value $\mu_1 + \cdots + \mu_{j-1} + i + 1$. Hence, the row reading tableau of shape $\mu = 421$ is that of Figure 5.2. As a general rule, a representation $\mathcal{V}$ of a group G is said to be *cyclic* if $\mathcal{V} = \mathbb{C}[\gamma \cdot v \mid \gamma \in G]$ for some element v of $\mathcal{V}$. In this sense, $\mathcal{H}^\mu$ is clearly cyclic. Among the special cases that are easy to characterize, we notice that $\mathcal{H}^{(n)} = \mathcal{L}[1] = \mathbb{C}$, on which $\mathfrak{S}_n$ acts trivially. For $\mu = 1^n$, whose cells are of the form $(0,j)$, the row reading tableau is simply $\tau(0,j) := j + 1$, hence $\mathbf{x}^\tau = x_2 x_3^2 x_4^3 \cdots x_n^{n-1}$. It follows that $\mathcal{H}^{(1^n)}$ is isomorphic to the left regular representation of $\mathfrak{S}_n$, whose dimension is $n!$. More generally, the dimension of $\mathcal{H}^\mu$ is $n!/(\mu_1! \cdots \mu_k!)$.

---

[2]Notice that this is not the evaluation monomial $\mathbf{x}_\tau$ of Section 4.1.

5.5   Irreducible Representations of $\mathfrak{S}_n$

Our aim now is to describe an explicit construction for the irreducible representations of $\mathfrak{S}_n$. Moreover, we intend to tie this construction to the cyclic structure (partition) describing a given conjugacy class of $\mathfrak{S}_n$. To this end, for a partition μ, we consider the polynomials

$$\Delta_\tau = \Delta_\tau(\mathbf{x}) := \sum_{\sigma \in \mathrm{Col}_\tau} \varepsilon(\sigma)\sigma \cdot \mathbf{x}^\tau, \qquad (5.5)$$

associated with a shape μ bijective tableau τ. Recall that $\varepsilon(\sigma)$ is the sign of σ. Calculating with the tableau τ of Figure 5.2 and factoring, we find that

$$\Delta_\tau = (x_1 - x_5)(x_1 - x_5)(x_5 - x_7)(x_2 - x_6).$$

In general, Δ_τ is the product of all possible factors $x_i - x_j$ with i appearing below j in a same column of τ. Clearly, we have $\sigma \cdot \Delta_\tau(\mathbf{x}) = \Delta_{\sigma \cdot \tau}(\mathbf{x})$, hence we can introduce the (cyclic) $\mathfrak{S}_n$-module

$$\mathcal{S}^\lambda := \mathcal{L}[\Delta_\tau(\mathbf{x}) \mid \tau \text{ of shape } \lambda], \qquad (5.6)$$

spanned by the polynomials $\Delta_\tau(\mathbf{x})$. A direct application of the definition gives that

$$\mathcal{S}^{21} = \mathcal{L}[x_3 - x_1, x_2 - x_1, x_3 - x_2].$$

This two-dimensional space affords the basis $\{x_3 - x_1, x_2 - x_1\}$. For a more general family of examples, consider the case when $\lambda = (n)$ for which we get the *trivial representation* $\mathcal{S}^{(n)} = \mathbb{C}$. At the other extreme, we get the *alternating representation* $\mathcal{S}^{(1^n)} = \mathcal{L}[\sigma \cdot \Delta_n(\mathbf{x}) \mid \sigma \in \mathfrak{S}_n]$, with $\Delta_n(\mathbf{x})$ coinciding with the classical Vandermonde determinant (see Section 3.9). Since $\Delta_n(\mathbf{x})$ is antisymmetric, we have $\dim \mathcal{S}^{(1^n)} = 1$. The general fact is that the set $\{\Delta_\tau(\mathbf{x}) \mid \tau \text{ standard tableau of shape } \lambda\}$ is always a basis of $\mathcal{S}^\lambda$, implying that

$$\dim \mathcal{S}^\lambda = f^\lambda.$$

Recall that f^λ is the number of standard tableaux of shape λ. For each μ, the space $\mathcal{S}^\mu$ is a G-invariant subspace of $\mathcal{H}^\mu$, and we have the decomposition

$$\mathcal{H}^\mu = \mathcal{S}^\mu \oplus \bigoplus_{\lambda \succ \mu} K_{\lambda,\mu}\mathcal{S}^\lambda. \qquad (5.7)$$

In fact there is a nontrivial representation homomorphism $f\colon \mathcal{S}^\lambda \longrightarrow \mathcal{H}^\mu$ if and only if μ is smaller or equal to λ in the dominance order. The theorem that describes all the situations is as follows.

Theorem 5.1 (Classical). *With λ and μ denoting partitions of n, we have the following:*

(1) *the $\mathfrak{S}_n$-modules $\mathcal{S}^\lambda$ are all irreducible;*

(2) *$\mathcal{S}^\lambda$ is isomorphic to $\mathcal{S}^\mu$ iff $\lambda = \mu$;*

(3) *$\{\mathcal{S}^\lambda\}_{\lambda \vdash n}$ is a complete system of irreducible representations of $\mathfrak{S}_n$;*

(4) *the Frobenius transform[3] of the character of $\mathcal{S}^\lambda$ is the Schur function s_λ.*

5.6   Frobenius Transform

A very nice feature of $\mathfrak{S}_n$ is that we can describe characters in terms of symmetric functions, rather like sequences of numbers can be described in terms of formal power series (generating series). This is achieved through the Frobenius transform, which is in fact defined for all central functions on $\mathfrak{S}_n$. We have already observed that characters are central functions. The *Frobenius transform* of a central function φ is defined as

$$\mathcal{F}(\varphi) := \frac{1}{n!} \sum_\sigma \varphi(\sigma) p_{\lambda(\sigma)}, \tag{5.8}$$

where $\lambda(\sigma)$ is the partition that gives the cyclic structure of σ (see Section 1.6). A word of caution may be in order here. In the sequel we often consider $\mathfrak{S}_n$-modules of polynomials, most often in the variables $\mathbf{x}$. The Frobenius transform of their character is expressed in terms of "formal" symmetric functions $p_\lambda = p_\lambda(\mathbf{z})$ whose variables $\mathbf{z} = z_1, z_2, \ldots$ have no link to the variables $\mathbf{x} = x_1, \ldots, x_n$. In this context the symmetric functions $f(\mathbf{z})$ only play the formal role of "markers", although we will see that they do this in a particularly efficient manner. Let us illustrate with the representation $\mathcal{S}^{(n-1)1}$ described in Section 5.5. The underlying space is $(n-1)$-dimensional and affords as basis the set

$$\{x_2 - x_1, x_3 - x_1, \ldots, x_n - x_1\}.$$

Exercise. Show that the value of the character $\chi^{(n-1)1}$ on a permutation σ is equal to the number of fixed points of σ minus 1.

It follows from this exercise that

$$\mathrm{Frob}(\mathcal{S}^{(n-1)1}) = \sum_{\lambda = 1^{d_1} \cdots \mathbf{n}^{d_n}} (d_1 - 1) \frac{p_1^{d_1} \cdots p_n^{d_n}}{1^{d_1} d_1! \cdots n^{d_n} d_n!}$$

$$= h_{n-1} h_1 - h_n = s_{(n-1)1},$$

---

[3] See Section 5.6.

with the last equality obtained from the Jacobi–Trudi formula. This goes to illustrate part (4) of Theorem 5.1. The last formula exhibits a general feature of the Frobenius transform of a central function. Indeed, summands in (5.8) are invariant under conjugation, i.e.,

$$\varphi(\tau^{-1}\sigma\tau)p_{\lambda(\tau^{-1}\sigma\tau)} = \varphi(\sigma)p_{\lambda(\sigma)}.$$

We can thus collect equal terms in the definition to get the equivalent expression

$$\mathcal{F}(\varphi) := \sum_{\mu \vdash n} \varphi(\sigma_\mu)\frac{p_\mu}{z_\mu}, \tag{5.9}$$

where σ_μ is any given permutation of shape μ. This last rewriting comes from the observation that $n!/z_\mu$ is the number of permutations of shape μ.

An evident basis of the space $\mathcal{C}(\mathfrak{S}_n)$ is given by the *class characteristic functions*:

$$C_\mu(\tau) := \begin{cases} 1 & \text{if } \lambda(\tau) = \mu, \\ 0 & \text{otherwise,} \end{cases}$$

one for each partition of n. One traditionally equips $\mathcal{C}(\mathfrak{S}_n)$ with the scalar product

$$\langle C_\lambda, C_\mu \rangle := \begin{cases} 1/z_\lambda & \text{if } \lambda = \mu, \\ 0 & \text{otherwise.} \end{cases}$$

This turns the Frobenius transform $\mathcal{F}: \mathcal{C}(\mathfrak{S}_n) \longrightarrow \mathbb{C}[\mathbf{z}]^{\mathfrak{S}_*}$ into a scalar product preserving linear map, since $\mathcal{F}(C_\mu) = p_\mu/z_\mu$.

For any $\mathfrak{S}_n$-module $\mathcal{V}$, we denote by $\mathcal{F}(\mathcal{V})$ (rather than $\mathcal{F}(\chi_\mathcal{V})$) the Frobenius transform of the character $\chi_\mathcal{V}$ of $\mathcal{V}$. In symbols, writing $\chi(\mu)$ for $\chi(\sigma)$ whenever $\lambda(\sigma) = \mu$, we get

$$\mathcal{F}(\mathcal{V}) = \sum_{\mu \vdash n} \chi_\mathcal{V}(\mu)\frac{p_\mu}{z_\mu}. \tag{5.10}$$

To simplify the terminology, let us say that $\mathcal{F}(\mathcal{V})$ is the *Frobenius characteristic* of $\mathcal{V}$. Part (4) of Theorem 5.1 can now be written as $\mathcal{F}(\mathcal{S}^\lambda) = s_\lambda$. Thus, we can say that the *character table* of the symmetric group $\mathfrak{S}_n$ describes the expansion of Schur functions in terms of the power sum basis.

Illustrating with $n = 4$, we transform the symmetric function identities

$$\mathcal{F}(\mathcal{S}^\lambda) = s_\lambda = \sum_{\mu \vdash n} \chi^\lambda(\mu)\frac{p_\mu}{z_\mu}$$

into the form of Table 5.1, denoting by $\chi^\lambda(\mu)$ the value of the character of an irreducible representation $\mathcal{S}^\lambda$ at permutations of shape μ. Observe that

$\lambda \backslash^{\mu}$	1111	211	31	22	4
4	1	1	1	1	1
31	3	1	-1	0	-1
22	2	0	2	-1	0
211	3	-1	-1	0	1
1111	1	-1	1	1	-1

Table 5.1. Character table of $\mathfrak{S}_4$.

we can deduce the dimension of the representation $\mathcal{V}$ from its Frobenius characteristic by the simple device of taking the scalar product with p_1^n. In formula:

$$\dim(\mathcal{V}) = \langle \mathcal{F}(\mathcal{V}), p_1^n \rangle, \tag{5.11}$$

since this is equal to the value of the character of $\mathcal{V}$ at the identity.

To get to the main point of this section, we need the further observation that $\mathcal{F}(\mathcal{V} \oplus \mathcal{W}) = \mathcal{F}(\mathcal{V}) + \mathcal{F}(\mathcal{W})$. This makes it clear that the irreducible decomposition of a $\mathfrak{S}_n$-module $\mathcal{V}$ is entirely described by the expansion of $\mathcal{F}(\mathcal{V})$ in terms of Schur functions. In symbols, we have

$$\mathcal{F}(\mathcal{V}) = \sum_{\lambda \vdash n} a_\lambda \mathcal{F}(\mathcal{V}^\lambda) = \sum_{\lambda \vdash n} a_\lambda s_\lambda, \tag{5.12}$$

where a_λ is the multiplicity of the irreducible $\mathfrak{S}_n$-module $\mathcal{V}^\lambda$ in $\mathcal{V}$. In particular, we see that any symmetric function arising as the Frobenius characteristic of some $\mathfrak{S}_n$-module is de facto Schur-positive. Let us illustrate this crucial point with a family of $\mathfrak{S}_n$-modules for which we have already obtained the irreducible decomposition. Indeed, in view of (4.17) and (5.7), we have

$$\mathcal{F}(\mathcal{H}^\mu) = \sum_\lambda K_{\lambda,\mu} s_\lambda = h_\mu. \tag{5.13}$$

Thus, we get a representation-theoretic explanation of identity (4.17), and of the fact that h_μ is Schur-positive.

Exercise. Expand the Frobenius characteristic of the defining representation of $\mathfrak{S}_n$ in terms of the complete homogeneous symmetric function basis.

The Regular Representation of $\mathfrak{S}_n$

Now that we have these tools at hand, it is easy to describe the Frobenius characteristic of the regular representation of $\mathfrak{S}_n$. This is worth exploring in some detail since the regular representation $\mathcal{R}$ appears in many guises in the sequel. A direct translation of our discussion in Sections 5.3 and

5.6 implies that the Frobenius characteristic of $\mathcal{R}$ is simply[4] h_1^n. Using our knowledge of symmetric function identities and (5.3), we find that

$$h_1^n = \sum_{\lambda \vdash n} f^\lambda s_\lambda, \qquad (5.14)$$

with f^λ equal to the number of standard Young tableaux of shape λ (given by the hook formula). In other words (as was already mentioned in (5.3)) each irreducible representation of $\mathfrak{S}_n$ does occur in $\mathcal{R}$, and in as many copies as its dimension. For small values of n, we have

$$h_1^2 = s_2 + s_{11},$$
$$h_1^3 = s_3 + 2s_{21} + s_{111},$$
$$h_1^4 = s_4 + 3s_{31} + 2s_{22} + 3s_{211} + s_{1111},$$
$$h_1^5 = s_5 + 4s_{41} + 5s_{32} + 6s_{311} + 5s_{221} + 4s_{2111} + s_{11111}.$$

Recall that the coefficients represent both the multiplicity and dimension of the associated irreducible representations.

The Conjugating Representation of $\mathfrak{S}_n$

We start with the observation that z_μ is the value of the character of the conjugating representation at a permutation of shape μ. Indeed, it is a basic exercise in combinatorics to show that z_μ is the number of elements of the set $\{\tau \mid \sigma^{-1}\tau\sigma = \tau\}$, if σ has a cycle structure given by the partition μ. It follows from the definition that the corresponding Frobenius characteristic F_n is equal to $\sum_{\mu \vdash n} p_\mu$. Expanding the resulting expressions in the Schur function basis gives a description of how this representation decomposes into irreducible ones.

For small values of n, we have the following expansions:

$$F_1 = s_1,$$
$$F_2 = 2s_2,$$
$$F_3 = 3s_3 + s_{21} + s_{111},$$
$$F_4 = 5s_4 + 2s_{31} + 3s_{22} + 2s_{211} + s_{1111},$$
$$F_5 = 7s_5 + 5s_{41} + 6s_{32} + 5s_{311} + 4s_{221} + 3s_{2111} + s_{11111},$$
$$F_6 = 11s_6 + 8s_{51} + 15s_{42} + 10s_{411} + 4s_{33} + 13s_{321} + 10s_{3111}$$
$$\qquad + 8s_{222} + 5s_{2211} + 4s_{21111} + s_{111111}.$$

Exercise. Show that the coefficient of s_n in F_n is the number of partitions of n.

---

[4]It could also be written as p_1^n, s_1^n or e_1^n, since $h_1 = p_1 = s_1 = e_1$.

5.7 Restriction from $\mathfrak{S}_n$ to $\mathfrak{S}_{n-1}$

By the general process of restriction of a representation of a group G to a subgroup H, we get representations of $\mathfrak{S}_{n-1}$ from representations of $\mathfrak{S}_n$. Here we identify $\mathfrak{S}_{n-1}$ with the subgroup of $\mathfrak{S}_n$ whose support is the set of permutations that fix n. The Frobenius characteristic of the restriction $\mathrm{Res}^n_{n-1}(\mathcal{V})$ of a $\mathfrak{S}_n$-representation $\mathcal{V}$ to this subgroup can easily be described as

$$\mathrm{Frob}_t\left(\mathrm{Res}^n_{n-1}(\mathcal{V})\right) = s_1^\perp \, \mathrm{Frob}_t(\mathcal{V}). \tag{5.15}$$

Indeed, writing χ for the character of $\mathcal{V}$ and $\chi|_{n-1}$ for the character of the restriction considered, we find that

$$\sum_{\nu \vdash n-1} \chi|_{n-1}(\nu)\frac{p_\nu}{z_\nu} = \sum_{\nu \vdash n-1} \chi(\nu \cup 1)\frac{p_\nu}{z_\nu}$$
$$= s_1^\perp \sum_{\mu \vdash n} \chi(\mu)\frac{p_\mu}{z_\mu},$$

since $s_1^\perp$ acts as a derivation with respect to the variable p_1 on the power sum basis. In particular, it follows from the Pieri rule (see Section 4.7) that the restriction to $\mathfrak{S}_{n-1}$ of the λ-indexed $\mathfrak{S}_n$-irreducible representation decomposes into a sum

$$\mathrm{Res}^n_{n-1}(\mathcal{V}^\lambda) \simeq \bigoplus_{\mu \to \lambda} \mathcal{V}^\mu$$

of μ-indexed $\mathfrak{S}_{n-1}$-irreducible representations over the set of partitions μ that can be obtained from λ by removing one corner.

5.8 Polynomial Representations of $\mathrm{GL}\,(\mathcal{V})$

From an elementary point of view, a *polynomial representation* of the group GL_n, of invertible $n \times n$ matrices over $\mathbb{C}$, is just a group homomorphism

$$\rho\colon \mathrm{GL}_n \longrightarrow \mathrm{GL}_N,$$

such that the entries of the matrix $\rho(M)$ are polynomials in the entries of the matrix M. For example, we can check that the map

$$\begin{pmatrix} a & b \\ c & d \end{pmatrix} \longmapsto \begin{pmatrix} a^2 & 2ab & b^2 \\ ac & ad+bc & bd \\ c^2 & 2cd & d^2 \end{pmatrix} \tag{5.16}$$

establishes such an homomorphism. It is also well known that $\det\colon \mathrm{GL}_n \to \mathbb{C}^*$ is a group homomorphism. It clearly gives another example of such a polynomial representation if we identify $\mathbb{C}$ with GL_1. Evidently all of this generalizes to $GL(\mathcal{V})$, for $\mathcal{V}$ an n-dimensional vector space. From this more intrinsic point of view, one of the basic constructions is that of the kth *tensorial power* $\mathcal{V}^{\otimes k}$. Recall that for B an ordered basis of $\mathcal{V}$, the vector space $\mathcal{V}^{\otimes k}$ affords as basis the set $B^{\otimes k}$ of tensors $v_1 \otimes \cdots \otimes v_k$, with the v_j varying freely in the set B. Thus the dimension of $\mathcal{V}^{\otimes k}$ is n^k. Using the multilinearity of the tensor product, any invertible linear map $T : \mathcal{V} \to \mathcal{V}$ naturally gives rise to an invertible linear map $T^{\otimes k} : \mathcal{V}^{\otimes k} \to \mathcal{V}^{\otimes k}$. If M is the matrix of the map T with respect to the basis B, then the matrix of $T^{\otimes k}$ in the basis $B^{\otimes k}$, ordered lexicographically, is an $n^k \times n^k$ matrix $M^{\otimes k}$ recursively obtained as follows. Respectively denote by a_{ij} and $b_{\ell m}$ the entries of M and $M^{\otimes k-1}$. For indices r and s between 1 and n^k, find the unique expressions of the form $r = (i-1)n^{k-1} + \ell$ and $s = (j-1)n^{k-1} + m$, with $1 \le i, j \le n$ and $1 \le \ell, m \le n^{k-1}$. We then set the corresponding entry of $M^{\otimes k}$ to be $c_{rs} := a_{ij}b_{\ell m}$. For example,

$$
\begin{pmatrix} a & b \\ c & d \end{pmatrix}^{\otimes 2} = \begin{pmatrix} a & b \\ c & d \end{pmatrix} \otimes \begin{pmatrix} a & b \\ c & d \end{pmatrix} = \begin{pmatrix} aa & ab & ba & bb \\ ac & ad & bc & bd \\ ca & cb & da & db \\ cc & cd & dc & dd \end{pmatrix}.
$$

The *character* of a polynomial representation ρ is defined as follows. We consider the diagonal matrix

$$
D(\mathbf{x}) = \begin{pmatrix} x_1 & 0 & \cdots & 0 \\ 0 & x_2 & \cdots & 0 \\ \vdots & \vdots & \ddots & \vdots \\ 0 & 0 & \cdots & x_n \end{pmatrix},
$$

with $x_1, \ldots, x_n$ on the main diagonal and 0 in every other entry. The character of ρ is then defined to be $\chi_\rho(\mathbf{x}) := \mathrm{Trace}\left(\rho(D(\mathbf{x}))\right)$. It is clearly a symmetric function of the $\mathbf{x} = x_1, \ldots, x_n$. Calculating the character of the representation $\mathcal{V}^{\otimes k}$, we get $h_1(\mathbf{x})^k = (x_1 + \cdots + x_n)^k$. For the representation in (5.16) we get $h_2(x_1, x_2) = x_1^2 + x_1 x_2 + x_2^2$. Another interesting general example corresponds to the representation associated with the determinant map. In this case, we get as a character the symmetric function $e_n(\mathbf{x}) = x_1 x_2 \cdots x_n$.

It may appear strange, at first glance, that we only compute the character for diagonal matrices. Recall, however, that the trace is a continuous function $\chi : GL_n \to \mathbb{C}$, which is constant on conjugacy classes. Moreover, the set of diagonalizable matrices is dense in the space of matrices. Thus, the choice of definition for the character function is seen to be natural, since it is essentially "defined" over a dense subset.

5.9   Schur–Weyl Duality

We can construct all polynomial representations of $GL(\mathcal{V})$ using a special case of a more general notion known as Schur–Weyl duality. We start by considering the action of $\mathfrak{S}_k$ on k-tensors in $\mathcal{V}^{\otimes k}$ by permutation of entries, i.e., $\sigma \cdot v_1 \otimes \cdots \otimes v_k = v_{\sigma(1)} \otimes \cdots \otimes v_{\sigma(k)}$. Then, given any linear representation $\mathcal{W}$ of $\mathfrak{S}_k$, we can consider the action of $GL(\mathcal{V})$ on the tensor product[5] $\mathcal{W} \otimes_\mathcal{R} \mathcal{V}^{\otimes k}$ over the ring $\mathcal{R} = \mathbb{C}\mathfrak{S}_k$ (known as the *group algebra* of $\mathfrak{S}_n$), with $GL(\mathcal{V})$ acting pointwise on the part $\mathcal{V}^{\otimes k}$. To be sure that this definition makes sense, we need to check the following.

Exercise. Show that the action of $\mathfrak{S}_k$ on $\mathcal{V}^{\otimes k}$ commutes with the action of $GL(\mathcal{V})$ on $\mathcal{V}^{\otimes k}$.

The fact is that all polynomial representations of $GL(\mathcal{V})$ arise in this manner. Moreover, the irreducible polynomial representations of $GL(\mathcal{V})$ correspond exactly to irreducible representations of $\mathfrak{S}_k$, with $1 \le k \le n$, having the same characters.

Rather than give systematic proofs, let us illustrate all this with an example. Consider $\mathcal{S}^\lambda$ with $\lambda = (k-1, 1)$, the irreducible representation of $\mathfrak{S}_k$ described in Section 5.6. As before let $\mathcal{V}$ be a vector space having basis B. For any vectors u in $\mathcal{S}^\lambda$ and v in $\mathcal{V}^{\otimes k}$, if σ is a permutation in $\mathfrak{S}_k$ then we have $(\sigma \cdot u) \otimes v = u \otimes (\sigma \cdot v)$ in the space $\mathcal{S}^\lambda \otimes_\mathcal{R} \mathcal{V}^{\otimes k}$. It follows that the set of tensors of the form $(x_n - x_1) \otimes v$ spans $\mathcal{S}^\lambda \otimes_\mathcal{R} \mathcal{V}^{\otimes k}$. It can be checked that the corresponding character is the Schur function $s_{(k-1)1^k}$.

Exercise. Construct a basis of $\mathcal{S}^\lambda \otimes_\mathcal{R} \mathcal{V}^{\otimes k}$ naturally indexed by semi-standard tableaux of shape $\lambda = (k-1, 1)$, with entries in $\{1, \ldots, n\}$.

Two important cases are the polynomial representations that correspond to the trivial and sign representations of $\mathfrak{S}_k$. The respective $GL(\mathcal{V})$ polynomial representations obtained are the symmetric kth-power $S^k\mathcal{V}$ and the exterior kth-power $\bigwedge^k \mathcal{V}$.

Restricting a $\mathbf{GL}_n$ Representation to $\mathfrak{S}_n$

Many interesting representations of $\mathfrak{S}_n$ can be obtained by restricting polynomial representations of GL_n to the subgroup $\mathfrak{S}_n$ of permutation matrices. We are going to illustrate this process considering a polynomial representation of GL_n which is obtained using Schur–Weyl duality from a representation of $\mathfrak{S}_k$. The resulting polynomial representation is then restricted to $\mathfrak{S}_n$.

---

[5]Be careful here; we are considering the tensor product over the group algebra of $\mathfrak{S}_n$. In general, this introduces some relations between the tensors in $\mathcal{V}^{\otimes k}$.

For our illustration, consider the trivial representation $\mathcal{S}^{(k)}$ of $\mathfrak{S}_n$ as a starting point. By Schur–Weyl duality we get a GL_n-module $\mathcal{S}^{(k)} \otimes_{\mathcal{R}}$ $(\mathbb{C}^n)^{\otimes k}$ that can be identified with the space of degree k homogeneous polynomials in the variables $x_1, \ldots, x_n$. Restricting to $\mathfrak{S}_n$, we get a representation whose "graded" Frobenius characteristic is as given below. This new notion of "graded" Frobenius characteristic is going to become natural in the coming chapters. For the moment, it may be considered as an efficient tool for the description of an infinite sequence of linked results. It takes the form of formal power series in an auxiliary variable t, with the coefficient of t^k giving the Frobenius characteristic of $\mathcal{S}^{(k)} \otimes_{\mathcal{R}} (\mathbb{C}^n)^{\otimes k}$. Then

with $n = 1$ we get $s_1/(1 - t)$,

with $n = 2$ we get $(s_2 + ts_{11})/\big((1 - t)(1 - t^2)\big)$,

with $n = 3$ we get $\big(s_3 + (t + t^2)s_{21} + t^3 s_{111}\big)/\big((1 - t)(1 - t^2)(1 - t^3)\big)$,

etc.

A more systematic description will be given in Section 7.3. Other representations giving interesting restrictions of the kind are described at the end of Chapter 6. If f and g are the respective Frobenius characteristics of the restriction to $\mathfrak{S}_n$ of G_n-modules F and G, then the respective Frobenius characteristic of $F \oplus G$ and $F \otimes G$ are $f + g$ and $f * g$ (Kronecker product).

Exercise. For a graded representation $\mathcal{W}$ of $\mathfrak{S}_n$, show that

$$\mathrm{Frob}_t(\mathcal{W})[\mathbf{z}(1 - t)] = \sum_{k \geq 0}(-1)^k \mathrm{Frob}_t(\mathcal{W} \otimes \bigwedge^k \mathcal{V})(\mathbf{z})t^k, \qquad (5.17)$$

$$\mathrm{Frob}_t(\mathcal{W})[\mathbf{z}/(1 - t)] = \sum_{k \geq 0} \mathrm{Frob}_t(\mathcal{W} \otimes S^k \mathcal{V})(\mathbf{z})t^k \qquad (5.18)$$

with $\bigwedge^k \mathcal{V}$ and $S^k \mathcal{V}$ respectively denoting the exterior and symmetric kth-powers of the defining representation V.

Chapter 6

A Short Introduction to the Theory of Species

Following a slow (and forced) evolution in the history of mathematics, the modern notion of function (due to Dirichlet, 1837) has been made independent of any actual description format. A similar process led André Joyal to introduce the notion of "species" in combinatorics [Joyal 81] to make the description of structures *independent of any specific format*. On one side, the theory serves as an elegant "explanation" for the surprising power of generating functions in the solution of structure enumeration. On another side, it makes clear and natural much of Pólya's theory for the enumeration of structures up to isomorphism. Moreover, species are naturally linked to the study of "polynomial functors", which give a classification of polynomial representations of the general linear group. We refer to [Bergeron et al. 98] for an in-depth reference to the theory of species. Our particular reason for recalling the basic notions of species is that it allows us to generalize many of the questions related to the central material of this book.

6.1   Species of Structures

Let $\mathbb{B}$ be the category of finite sets with bijections. A *species (of structures)* is simply a functor $F \colon \mathbb{B} \longrightarrow \mathbb{B}$. More explicitly, for each finite set A, we are given a finite set $F[A]$ whose elements are said to be the *structures* of species F on the underlying set A. Moreover, for each bijection $\varphi \colon A \to B$, we also have a bijection $F[\varphi] \colon F[A] \longrightarrow F[B]$, called the *transport of F-structures along* φ. The fact that F is functorial means that we further impose the conditions $F[\mathrm{Id}_A] = \mathrm{Id}_{F[A]}$, and $F[\psi \circ \varphi] = F[\psi]F[\varphi]$. We simply write $F = G$ whenever there exists an invertible natural transformation from

F to G. This is to say that, for each finite set A, there is a *(natural)* *bijection* θ_A between the sets $F[A]$ and $G[A]$. Naturality means that we have compatibility with transport of structures, i.e., $G[\varphi] \circ \theta_A = \theta_B \circ F[\varphi]$ for all bijections φ from A to B.

Another interpretation of the functoriality of F is that we have a family of compatible actions

$$\mathfrak{S}_A \times F[A] \longrightarrow F[A]$$

of the groups $\mathfrak{S}_A$ on the sets of F-structures on finite sets A. In action-like notation, we have $\sigma \cdot t := F[\sigma](t)$ for $t \in F[A]$. Orbits under this action are called *types of structures*. A type is an equivalence class for the relation "$\sim$" on $F[A]$, which is defined by setting $s \sim t$ if and only if $F[\sigma](s) = t$, for some σ in $\mathfrak{S}_A$. We then say that s and t are *isomorphic structures*, or that they have the same *type*.

Examples of Species

To connect this definition with the usual structures of combinatorics, a few examples are in order. These are all straightforward reformulations, in the language of species, of classical constructions of set theory or basic combinatorics. At the same time, we are also setting up notation. We have:

(1) The species $\mathcal{P}$ of *set partitions*. An element π of $\mathcal{P}[A]$ is a partition of the set A. Thus, π is a family of disjoint subsets of A, such that $A = \bigcup_{C \in \pi} C$. The transport of a partition π along $\varphi \colon A \xrightarrow{\sim} B$ is the partition $\{\varphi(C)\}_{C \in \pi}$.

(2) For any positive integer k, the species of *k-tuples*. This is the species for which $F[A] = A^k$, i.e., its structures are k-tuples of elements of A. For $\varphi \colon A \xrightarrow{\sim} B$, we set

$$\varphi^k(a_1, \ldots, a_k) = \big(\varphi(a_1), \ldots, \varphi(a_k)\big).$$

(3) The *power set* species $\wp$, is defined as $\wp[A] := \{B \mid B \subseteq A\}$, with $\wp[\varphi](B) = \varphi(B)$.

(4) The species $\mathcal{G}$ of *directed graphs* is such that $\mathcal{G}[A] = \wp[A \times A]$ with $\mathcal{G}[\varphi] = \wp[\varphi \times \varphi]$. Elements of $A \times A$ are potential *arcs* of the graph.

(5) The species $\mathfrak{S}$ of *permutations* is defined by

$$\mathfrak{S}_A = \mathfrak{S}[A] = \{\sigma \mid \sigma \colon A \xrightarrow{\sim} A, \text{ a bijection}\},$$

with conjugation as transport of structures: $\mathfrak{S}[\varphi](\sigma) = \varphi \sigma \varphi^{-1}$.

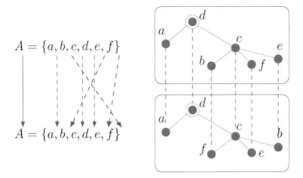

Figure 6.1. Transport as a relabeling of vertices.

(6) More generally, for the species End of *endofunctions*, we set

$$\text{End}[A] = \{f \mid f \colon A \to A\},$$

and again we consider conjugation as transport of structures:

$$\text{End}[\varphi](f) = \varphi f \varphi^{-1}.$$

In most of these constructions, the transport of structures can be viewed as a relabeling of the "vertices" as illustrated in Figure 6.1. When this is the case, transport of structures can easily be figured out as soon as we define the structures of the species considered.

6.2   Generating Series

The fact that the number of elements of $F[A]$ depends only on the cardinality of A is built into the definition of species. Indeed, we have $F[A] \xrightarrow{\sim} F[B]$ when $A \xrightarrow{\sim} B$. Let us denote by $\mathbf{n}$ the set $\{1, \dots, n\}$. To each species F we associate the *exponential generating series* $F(\zeta) := \sum_{n \geq 0} f_n \zeta^n / n$ where f_n is the number of elements of $F[\mathbf{n}]$. Thus, the formal power series $F(\zeta)$ contains all the information about the basic enumeration of F-structures. We compute generating series either directly using the definition, or as a solution of equations explicitly linked to combinatorial decompositions of the structures to be enumerated. This will be made clear in Section 6.3. Let us first directly compute a few series while introducing some other interesting species. We denote simply by X the species:

$$X[A] := \begin{cases} \{A\} & \text{if } A = 1, \\ \emptyset & \text{otherwise,} \end{cases}$$

which is "characteristic" of *singletons*. Its series is $X(\zeta) = \zeta$. Our next example is the rather simple species having exactly one structure on any set, i.e., $E[A] := \{A\}$. We say that E is the species of *sets*, and we have

$$E(\zeta) = \sum_{n \geq 0} \frac{\zeta^n}{n!} = e^\zeta = \exp(\zeta). \tag{6.1}$$

Since there are $n!$ permutations of an n-element set, we get (1) in the list below, which includes other simple examples (some of which have already been encountered):

(1) $\mathfrak{S}(\zeta) = \dfrac{1}{1 - \zeta}$, (2) $\wp(\zeta) = \exp(2\zeta)$, (3) $\mathcal{G}(\zeta) = \sum_{n \geq 0} 2^{n^2} \dfrac{\zeta^n}{n!}$,

(4) $\text{End}(\zeta) = \sum_{n \geq 0} n^n \dfrac{\zeta^n}{n!}$, (5) $\mathcal{C}(\zeta) = \log \dfrac{1}{1 - \zeta}$, (6) $\mathcal{E}(\zeta) = \zeta \exp(\zeta)$.

Here $\mathcal{C}$ stands for the species of *cyclic permutations*. It is a well-known fact that on an n-element set there are $(n - 1)!$ such permutations. Thus, we get (5). The species $\mathcal{E}$ of *elements* is defined by $\mathcal{E}[A] := A$, so that we have n structures on n-sets, hence we get (6).

6.3 The Calculus of Species

The theory of species really starts to get exciting with the introduction of operations that have nice properties. Among others, we have the operations of *sum*, *product*, *substitution*, and *derivative* of species. The interesting feature of these operations is that we have compatibility with the corresponding operations on generating series. Indeed, anticipating the upcoming definitions, we have the following equalities:

$$(F + G)(\zeta) = F(\zeta) + G(\zeta), \tag{6.2}$$

$$(F \cdot G)(\zeta) = F(\zeta)G(\zeta), \tag{6.3}$$

$$(F \circ G)(\zeta) = F\big(G(\zeta)\big), \tag{6.4}$$

$$F'(\zeta) = \frac{d}{d\zeta} F(\zeta). \tag{6.5}$$

Familiar identities also hold, such as $1 \cdot F = F$, $F \cdot (G + H) = F \cdot G + F \cdot H$, or the chain rule $(F \cdot G)' = F' \cdot G + F \cdot G'$. Here the species 1 is the *characteristic species for empty sets* defined by

$$1[A] = \begin{cases} \{A\} & \text{if } A = \emptyset, \\ \emptyset & \text{otherwise.} \end{cases}$$

More generally we have the *characteristic species for k-sets*:

$$E_k[A] = \begin{cases} \{A\} & \text{if } |A| = k, \\ \emptyset & \text{otherwise,} \end{cases}$$

whose series is $\zeta^k/k!$. We are going to describe a combinatorial calculus entirely parallel to the classical calculus of power series. It seems that most classical identities involving special functions can be "combinatorialized" in this manner. This is most certainly true of identities and formulas involving the classical families of orthogonal polynomials: *Hermite, Chebicheff, Laguerre, Jacobi*, etc. This has been done in a series of papers: [Bergeron 90, Foata 84, Foata and Leroux 83, Labelle and Yeh 89, Leroux and Strehl 85].

Formal Definitions

Here are the technical descriptions for the operations, followed below by illustrations that are almost as rigorous. For two species F and G, we introduce the species $F + G$, $F \cdot G$, $F \circ G$, and F', setting respectively

(1) $(F + G)[A] := F[A] + G[A]$.

(2) $(F \cdot G)[A] := \displaystyle\sum_{B+C=A} F[B] \times G[C]$.

(3) $(F \circ G)[A] := \displaystyle\sum_{\pi \in \mathcal{P}[A]} F[\pi] \times \prod_{B \in \pi} G[B]$, under the condition that $G[\emptyset] = \emptyset$.

(4) $F'[A] := F[A + \{*\}]$, with $\{*\}$ standing for any one-element set.

The "+" and "Σ", appearing on the right-hand sides of these definitions, are to be interpreted as *disjoint unions*. The associated transport of structures are all so evident that they need no special consideration. We will exploit these operations to construct new species out of known ones, and to write down equations. As a first simple illustration, let us consider the species X^n, recursively defined by

$$X^n := \begin{cases} X \cdot X^{n-1} & \text{if } n \geq 1, \\ 1 & \text{if } n = 0. \end{cases}$$

Its structures on A, correspond to all the possible ways of listing the elements of A, but only for sets of cardinality n. For other sets there are no structures. The next step may then be to construct the species of *lists*, $\mathcal{L} = \sum_{n \geq 0} X^n$, thus removing the condition on cardinality.

Exercise. Show that the infinite summation used in the definition of $\mathcal{L}$ makes sense. Check that

$$\mathcal{L}(\zeta) = \mathfrak{S}(\zeta). \tag{6.6}$$

Prove that the species $\mathcal{L}$ and $\mathfrak{S}$ are not isomorphic.

Typical Structures and Operations

Since the very beginning of the theory of species it has been clear that drawing the "right kind" of figure plays a crucial role in understanding intricate manipulations of operations. In the brief presentation outlined here, we should think of the equalities as describing "evident" bijective transformations. Giving too formal a description of these manipulations would necessitate going back to the original set-theoretical description, entirely defeating our objective. In each use of a drawing, we should ponder both sides of an equality, coming back to the following lexicon to decode portions of the manipulations. After an initial period of adaptation, we learn to read off formulas straight from the pictures.

The approach here consists of drawing "typical" structures of a species F, instead of writing $t \in F[A]$. If no property of F is known, the figure is very sketchy, as in Figure 6.2. Here the red points (with their accompanying edges) stand for the elements of the underlying set, and the F-labeled arc stands for some F-construction on these elements. Even with such a sketchy presentation, the product and substitution of species can be nicely presented as follows. As illustrated in Figure 6.3, a typical structure of

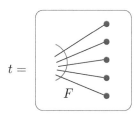

$$t =$$

Figure 6.2. A typical F-structure.

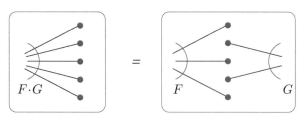

Figure 6.3. A typical $(F \cdot G)$-structure.

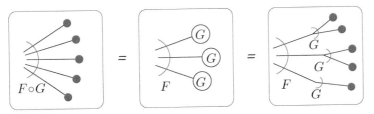

Figure 6.4. A typical $(F \circ G)$-structure.

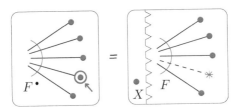

Figure 6.5. A typical $F^{\bullet}$-structure.

species $F \cdot G$ is obtained by splitting the underlying set into two parts; on the "left side" we choose a typical F-structure, and on the "right side" a typical G-structure.

Similarly, we represent typical structures of species $F \circ G$, also denoted by $F(G)$, as illustrated in Figure 6.4. One begins by choosing some partition of the underlying set. On each part (represented by big circles in the figure below) we choose a G-structure, and on the set of parts we choose an F-structure. The resulting structure is said to be an "F of G" structure. Among the many operations that can be constructed out of these basic operations, let us at least mention that of pointing. The species $F^{\bullet} = X \cdot F'$ of *pointed* F-structures has typical structure as illustrated in Figure 6.5. If f_n is the number of F-structures on $\mathbf{n}$, then nf_n is the number of $F^{\bullet}$-structures on $\mathbf{n}$, since an F-structure can be made into a pointed F-structure by selecting any of the n underlying points.

Other important examples of new operations are obtained by substitutions into fixed species such as E, $\mathfrak{S}$ or $\mathcal{L}$. For instance, the first of these, mapping F to $E(F)$, gives rise the species of sets of F-structures. This is expanded upon below.

Using the Calculus of Species

We may now begin using operations to set up identities. Recall that equality involves natural isomorphism of species, so that identities correspond to natural bijections between different outlooks on given structures. In other words, an identity is essentially a description of how to go from some

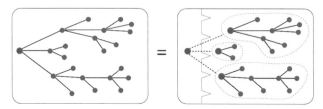

Figure 6.6. Canonical rooted tree decomposition corresponding to $\mathcal{A} = X \cdot E(\mathcal{A})$.

natural combinatorial decomposition of the structures involved to another. Moreover, each identity gives rise to an identity between associated generating series. This gives a powerful tool for the solution of combinatorial enumeration problems. Operations also allow the explicit or implicit construction of new species, with implicit constructions describing species as the "unique solution" of some equation in the algebra of species.

As we see below, basic identities are often "self evident" or "classical" if we read them out in the right manner. Examples of such identities are

(1) $E = 1 + E^+$: "a set is either empty, or nonempty";

(2) $\mathcal{P} = E(E^+)$: "a partition is a set of nonempty parts";

(3) $\mathfrak{S} = E(\mathcal{C})$: "a permutation is a set of disjoint cycles";

(4) $\mathcal{A} = X \cdot E(\mathcal{A})$, "a rooted tree is obtained by attaching to the *root* (an element of A) a set of rooted trees (the *branches*)".

A few comments are in order. The first identity is essentially a definition of the species E^+ of *nonempty sets*. The operation "+" is to be read as "or". The last identity is certainly the most interesting one. It implicitly defines the species of *rooted trees* $\mathcal{A}$ in a somewhat recursive form. This is based on a canonical decomposition of rooted trees as is described in Figure 6.6. The defining equation $\mathcal{A} = X \cdot E(\mathcal{A})$ for the rooted trees species $\mathcal{A}$ is an archetype of the functional equations that appear in general *Lagrange inversion* situations. By rewriting $f(\zeta)$ in the form[1] $\zeta/F(\zeta)$, we see that finding the composition inverse $g(\zeta)$ of a series $f(\zeta)$ is equivalent to solving

$$g(\zeta) = \zeta \, F\big(g(\zeta)\big). \tag{6.7}$$

All well-formed equations of this kind can be solved in the context of species. In Section 6.5 we solve a generic version of Equation 6.7.

A less well known, but still very classical, decomposition is that of endofunctions as permutations of rooted trees, as represented in Figure 6.7. From this we get the identity End $= \mathfrak{S}(\mathcal{A})$.

---

[1]We assume that $f(\zeta) = \zeta + a_2\zeta^2 + a_3\zeta^3 + \cdots$.

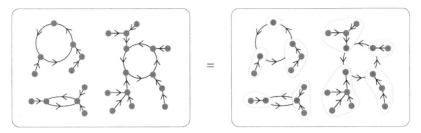

Figure 6.7. Canonical endofunction decomposition corresponding to End $= \mathfrak{S}(\mathcal{A})$.

6.4 Vertebrates and Rooted Trees

Passing to series, we get the equation $\mathcal{A}(\zeta) = \zeta \exp\big(\mathcal{A}(\zeta)\big)$ directly from the defining equation for $\mathcal{A}$, the species of trees. Solving this equation for the formal series $\mathcal{A}(\zeta) = \sum_{n\geq 0} a_n \zeta^n/n!$, we find that

$$\mathcal{A}(\zeta) = \zeta + 2\frac{\zeta^2}{2} + 9\frac{\zeta^3}{3!} + 64\frac{\zeta^4}{4!} + 625\frac{\zeta^5}{5!} + 7776\frac{\zeta^6}{6!} + 117649\frac{\zeta^7}{7!} + 2097152\frac{\zeta^8}{8!} + \cdots .$$

As seen below, the coefficient of $\zeta^n/n!$ is $a_n = n^{n-1}$. We show this following a very nice proof due to [Joyal 81]. More precisely, we check that $na_n = n^n$. Recall that n^n is the number of endofunctions on $\mathbf{n}$. We can thus exploit the following almost entirely combinatorial relation between the species $\mathcal{A}$ and End. In view of equation (6.6) and the identity corresponding to Figure 6.7, we have the series identities

$$\mathcal{L}(\mathcal{A})(\zeta) = \mathfrak{S}(\mathcal{A})(\zeta) = \text{End}(\zeta).$$

The rest of the argument boils down to verifying the species identity $\mathcal{A}^{\bullet} = \mathcal{L}(\mathcal{A})$. Indeed, this immediately implies that we have $na_n = n^n$, as announced. To get the required species identity, we consider the sequence of steps illustrated in Figure 6.8. This sequence of transformations turns a pointed rooted tree into an ordered list of rooted trees, as required. We begin with a pointed rooted tree. The central figure highlights the

Figure 6.8. Decomposition of a vertebrate as a list of rooted trees.

unique path going from the root to the selected point. This is the *vertebral column* of the "vertebrate". In view of this decomposition, we say that $\mathcal{A}^{\bullet}$ is the species of *vertebrates*. The *vertebra* are the points that lie on this vertebral column. To each vertebra is attached a rooted tree. In the last part of Figure 6.8, we see a vertebrate turned into a list of rooted trees.

6.5   Generic Lagrange Inversion

Drawing rooted trees in the plane can be somewhat misleading. Indeed, there are no actual differences between the two rooted trees (as structures of species $\mathcal{A}$) of Figure 6.9 (although we have to stare at them for a while to become convinced of this). The situation is quite different with the species $\mathcal{T}$ of *planar rooted trees* whose defining equation is $\mathcal{T} = X \cdot \mathcal{L}(\mathcal{A})$. In this case, the order of appearance of the branches that are attached to a node is an intricate part of the structure. Another way of underlining this difference is to observe that rooted trees often have *automorphisms*, i.e., permutations $\sigma \in \mathfrak{S}_A$ such that $\sigma \cdot t = t$, for $t \in \mathcal{A}[A]$, whereas planar rooted trees have no nontrivial automorphisms.[2] This fact implies that the number of planar rooted trees is a multiple of $n!$. Indeed, this number is equal to $n! \mathcal{C}_n$, with $\mathcal{C}_n$ standing for the Catalan number. This can be checked directly by solving the series equation associated with the defining equation for planar rooted trees.

As discussed in [Bergeron et al. 98], the notion of species can be extended in many directions. One of these allows the introduction of structures with weights lying in some adequate ring. In this manner we can easily understand the combinatorics behind the series solution of the following series equation. Most of the necessary ingredients have been outlined here,

---

[2]The identity is always an automorphism.

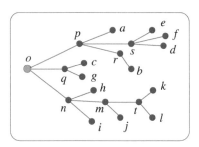
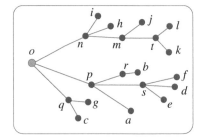

Figure 6.9. Two equal rooted trees.

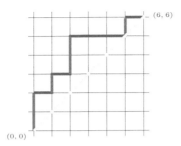

Figure 6.10. A weight $(2, 2, 1, 1)$ Dyck path.

but a complete presentation would become a bit too long. Let us just state the result that we will extend in Section 10.6. We want to compute the Lagrange inverse of the series

$$f(\zeta) = \zeta \Big(\sum_{n \geq 0} h_n \zeta^n \Big)^{-1}, \qquad (6.8)$$

with h_n the complete homogeneous symmetric functions in an infinite number of variables. This is the series $g(\zeta)$ such that $g(\zeta) = \zeta H\big(g(\zeta)\big)$, with $H(\zeta) = 1 + \sum_{k \geq 1} h_k \zeta^k$. The solution is

$$g(\zeta) = \zeta + h_1 \zeta^2 + (h_1^2 + h_2)\zeta^3 + (h_1^3 + 3h_2 h_1 + h_3)\zeta^4 + \cdots \qquad (6.9)$$

where the coefficient of ζ^{n+1} is given by the formula

$$g_n = \sum_{\gamma \in \mathcal{D}_n} h_{\omega(\gamma)}. \qquad (6.10)$$

Here, the sum is over the set $\mathcal{D}_n$ of length n "Dyck paths", with each path having weights $h_{\omega(\gamma)}$. To a path γ we associate a partition $\omega(\gamma)$ as defined below. It is convenient, from now on, to sightly modify our previous notion (see Section 1.7) of a Dyck path. The paths go from $(0, 0)$ to (n, n), and stay above the diagonal as illustrated in Figure 6.10. For a path γ, we consider the partition $\omega(\gamma)$ whose parts are the lengths of vertical segments of γ. The reason for considering equation (6.10) as a symmetric function is because it corresponds to the Frobenius characteristic of the action of the symmetric group on the span of the "parking function". We refer to [Garsia and Haiman 96a] for more on this.

As we have already mentioned in Section 3.5, the h_n are algebraically independent. We can thus specialize the h_n to any specified value. This is why we may think of (6.9) as giving a generic solution to Lagrange inversion.

6.6 Tensorial Species

A *tensorial species* is a functor $F\colon \mathbb{B} \longrightarrow \mathbb{V}$, where $\mathbb{V}$ is the category of (finite dimensional) vector spaces over $\mathbb{C}$. This corresponds essentially to a (compatible) family of *linear representations* $\mathfrak{S}_A \times F[A] \longrightarrow F[A]$ obtained by replacing sets by vector spaces in the preceding discussion, and bijection by invertible linear transform actions. Every species is naturally turned into a tensorial species by considering the free vector spaces spanned by its structures. We use the same notation for a species and the corresponding tensorial species. Clearly, we can linearize transport of structures for species to turn them into transport of structures for tensorial species.

Operations on tensorial species are defined just as on species, simply replacing disjoint unions by direct sums, and cartesian products by tensorial products. Hence, we have:

(1) $(F + G)[A] := F[A] \oplus G[A]$.

(2) $(F \cdot G)[A] := \displaystyle\bigoplus_{B+C=A} F[B] \otimes G[C]$.

(3) $(F \circ G)[A] := \displaystyle\bigoplus_{\pi \in \mathcal{P}[A]} F[\pi] \otimes \bigotimes_{B \in \pi} G[B]$, with the condition that $G[\emptyset] = \emptyset$.

(4) $F'[A] := F[A + \{*\}]$.

The associated generating series $F(\zeta) := \sum_{n \geq 0} \dim F[A] \zeta^n / n!$, are compatible with these operations. We can translate all manipulations on species into (almost) equivalent manipulations on the associated tensorial species. However, below we will broaden our point of view by passing to the tensorial context.

Examples of Tensorial Species

As a first example, let us consider the one-dimensional space[3]

$$\varepsilon[A] := \mathbb{C}\{ \bigwedge_{a \in A} a \}, \tag{6.11}$$

with the usual *wedge product*, i.e., such that $a \wedge b = -b \wedge a$. Since $\bigwedge_{a \in A} a$ is a basis of $\varepsilon[A]$, we can define $\varepsilon[\varphi]$, for $\varphi\colon A \to B$, by setting

$$\varepsilon[\varphi]\left(\bigwedge_{a \in A} a \right) := \bigwedge_{a \in A} \varphi(a). \tag{6.12}$$

---

[3]In principle, we need some order on A to define this, but the choice of order does not change the end result.

Another interesting case corresponds to a species reformulation of the description of irreducible representations of symmetric groups given in Section 5.5. For any partition λ of $n \geq 0$, we let $S^\lambda[A]$ be the vector space spanned by the polynomials $\Delta_\tau(\mathbf{x})$, for τ varying in the set of bijective A-fillings of λ. To this end we introduce variables x_a for each element of the set A and define tableau monomials just as before.

Frobenius Characteristic of Tensorial Species

Observe that for all finite sets A and tensorial species F, the group $\mathfrak{S}_A$ acts linearly on the space $F[A]$ by functoriality of F. Thus, we can introduce the *Frobenius characteristic series*

$$\mathrm{Frob}(F) := \sum_{n \geq 0} \frac{1}{n!} \sum_{\sigma \in \mathfrak{S}_n} \chi_{F[\mathbf{n}]}(\sigma) p_{\lambda(\sigma)}(\mathbf{z}). \tag{6.13}$$

In other words, $\mathrm{Frob}(F) := \sum_n \mathrm{Frob}(F[\mathbf{n}])$. In the context of species, the corresponding notion goes under the name of *cycle index series* (see [Bergeron et al. 98]). The important properties of Frobenius characteristic series are that we have

$$\mathrm{Frob}(F + G) = \mathrm{Frob}(F) + \mathrm{Frob}(G), \tag{6.14}$$

$$\mathrm{Frob}(F \cdot G) = \mathrm{Frob}(F)\,\mathrm{Frob}(G), \tag{6.15}$$

$$\mathrm{Frob}(F \circ G) = \mathrm{Frob}(F)[\mathrm{Frob}(G)], \tag{6.16}$$

where the operation on the right-hand side of (6.16) corresponds to plethysm. Recall that this corresponds to a bilinear multiplicative extension of the rules $p_k[p_j] = p_{kj}$, for all j, k. Special cases of Frobenius characteristic series are

(1) $\mathrm{Frob}(X) = p_1,$ (2) $\mathrm{Frob}(\mathcal{L}) = \dfrac{1}{1 - p_1},$

(3) $\mathrm{Frob}(\mathfrak{S}) = \displaystyle\prod_{k \geq 1} \frac{1}{1 - p_k},$ (4) $\mathrm{Frob}(E) = H,$

(5) $\mathrm{Frob}(\mathcal{P}) = H[H - 1],$ (6) $\mathrm{Frob}(\mathcal{C}) = \displaystyle\sum_{k=1}^\infty \frac{\varphi(k)}{k} \log \frac{1}{1 - p_k},$

$$\tag{6.17}$$

with $H = \sum_{n \geq 0} h_n$. Observe that the degree n homogeneous component of (3) corresponds to the Frobenius characteristic of the conjugating representation of $\mathfrak{S}_n$. For the Frobenius characteristic of the species of cyclic permutations, (6) is obtained using the identity $\mathfrak{S} = E(\mathcal{C})$, Möbius inversion, and some calculation. This is why the resulting expression makes use of Euler's φ-function. We get

$$\text{Frob}(\mathcal{C}) = s_1 + s_2 + (s_3 + s_{111}) + (s_4 + s_{22} + s_{211})$$
$$+ (s_5 + s_{32} + 2s_{311} + s_{221} + s_{11111}) + \cdots.$$

We also derive the following equation from the defining equation of rooted trees:

$$\text{Frob}(\mathcal{A}) = p_1 H[\text{Frob}(\mathcal{A})].$$

One finds that

$$\text{Frob}(\mathcal{A}) = h_1 + h_1^2 + (h_1 h_2 + h_1^3) + (h_1 h_3 + h_1^2 h_2 + 2h_1^4)$$
$$+ (h_1 h_4 + 2h_1 h_2^2 + h_1^2 h_3 + h_1^3 h_2 + 4h_1^5) + \cdots$$
$$= s_1 + (s_2 + s_{11}) + (2s_3 + 3s_{21} + s_{111})$$
$$+ (4s_4 + 9s_{31} + 5s_{22} + 7s_{211} + 2s_{1111})$$
$$+ (9s_5 + 26s_{41} + 28s_{32} + 30s_{311}$$
$$+ 24s_{221} + 17s_{2111} + 4s_{11111}) + \cdots.$$

As another illustration of the use of operations to easily derive Frobenius characteristic formulas, consider the species Der of *derangements*. Recall that derangements are permutations with no fixed points. One readily checks the species identity $\mathfrak{S} = E \cdot \text{Der}$. It follows that

$$\text{Frob}(\text{Der}) = H^{-1} \prod_{k \geq 1} \frac{1}{1 - p_k}$$
$$= 1 + 2s_1 + (4s_2 + s_{11}) + (7s_3 + 4s_{21} + s_{111})$$
$$+ (12s_4 + 9s_{31} + 6s_{22} + 4s_{211} + 2s_{1111})$$
$$+ (19s_5 + 19s_{41} + 17s_{32} + 11s_{311}$$
$$+ 10s_{221} + 7s_{2111} + 2s_{11111}) + \cdots.$$

Specializations

It should be noted that we can derive the generating series $F(\zeta)$ from the Frob(F) by the simple device of specializing the p_k, setting $p_1 = \zeta$ and $p_k = 0$ for $k \geq 2$. Observe that this sends the Schur function s_μ to $f^\mu \zeta^n / n!$.

Another classical specialization consists of setting $p_k = \zeta^k$. It transforms the Frobenius characteristic series of a species into the *type enumerator series* $\widetilde{F}(\zeta)$. Indeed, using Burnside lemma we find that

$$\widetilde{F}(\zeta) = \sum_{n \geq 0} \widetilde{f}_n \zeta^n,$$

with $\widetilde{f}_n$ counting the number of "orbits" of the action of $\mathfrak{S}_n$ on the set $F[\mathbf{n}]$. From the point of view of tensorial species, we interpret $\widetilde{f}_n$ as giving the coefficient of s_n in $\mathrm{Frob}(F)$. To conciliate the two approaches, we can think of each orbit as giving rise to a subspace affording the trivial representation. This subspace is simply the set of multiples of the sum of the elements in the orbit.

6.7 Polynomial Functors

The *polynomial functor* F associated with a tensorial species, also denoted by F, is the functor[4] $F \colon \mathbb{V} \longrightarrow \mathbb{V}_\infty$, defined as

$$F(V) := \sum_{n \geq 0} F[\mathbf{n}] \otimes_{\mathbb{C}\mathfrak{S}_n} V^{\otimes n}, \tag{6.18}$$

for V a finite dimensional vector space, with evident effect on linear isomorphisms. The $\mathbb{C}\mathfrak{S}_n$-module structure on $V^{\otimes n}$ corresponds to permutation of components. In a manner consistent with our previous definitions, we say that the *Hilbert series* of $F(V)$ is

$$\mathrm{Hilb}_q\big(F(V)\big) := \sum_{n \geq 0} q^n \dim(F[\mathbf{n}] \otimes_{\mathbb{C}\mathfrak{S}_n} V^{\otimes n}). \tag{6.19}$$

Examples of Polynomial Functors

The classical *tensor algebra* $T(V) := \sum_{n \geq 0} V^{\otimes n}$, is obtained by applying the polynomial functor $\mathcal{L}$ on V.

Exercise. Check that $\mathcal{L}[\mathbf{n}] \otimes_{\mathbb{C}\mathfrak{S}_n} V^{\otimes n}$ and $V^{\otimes n}$ are isomorphic as $\mathrm{GL}(V)$-modules. Given a basis $\{x_1, \ldots, x_n\}$ of V, show that $T(V)$ identifies with the ring of polynomials in the noncommutative variables x_i. Calculate the associated Hilbert series.

The *exterior algebra* $\bigwedge(V) := \sum_{n \geq 0} \bigwedge^n V$, corresponds to the ε tensor species, since we have the natural isomorphism $\varepsilon[\mathbf{n}] \otimes_{\mathbb{C}\mathfrak{S}_n} V^{\otimes n} \simeq \bigwedge^n V$.

Exercise. Given a basis $\{x_1, \ldots, x_n\}$ of V, show that the vector space $\bigwedge(V)$ affords as generators the vectors $x_{i_1} \wedge x_{i_2} \wedge \cdots \wedge x_{i_k}$, with $\{i_1, i_2, \ldots, i_k\}$ varying in the subsets of $\{1, \ldots, n\}$. Show also that

$$x_{\sigma(i_1)} \wedge \cdots \wedge x_{\sigma(i_k)} = \varepsilon(\sigma)\, x_{i_1} \wedge x_{i_2} \wedge \cdots \wedge x_{i_k}$$

for all permutations σ of the set $\{i_1, i_2, \ldots, i_k\}$. Give a formula for the Hilbert series of $\bigwedge(V)$ and calculate the associated Hilbert series.

---

[4]With value in the category of (graded) vector spaces. These need not be of finite dimension, but each homogeneous part $F[\mathbf{n}] \otimes_{\mathbb{C}\mathfrak{S}_n} V^{\otimes n}$ is of finite dimension.

The *symmetric algebra* $S(V) := \sum_{n \geq 0} S^n(V)$ corresponds to the species E of sets, since $E[\mathbf{n}] \otimes_{\mathbb{C}\mathfrak{S}_n} V^{\otimes n} \simeq S^n(V)$.

Exercise. Given a basis $\{x_1, \ldots, x_n\}$ of V, show that $S(V)$ identifies with the ring of polynomials in the commutative variables x_i. Calculate the associated Hilbert series.

Polynomial functors associated with interesting species often correspond to nice $GL(V)$ representations. In turn, these can be restricted to $\mathfrak{S}_n$ (when V is n dimensional) naturally giving rise to spaces for which questions can be raised that are addressed in following chapters. For more on these notions see [Joyal 86].

Chapter 7

Some Commutative Algebra

This chapter is intended to be a fast-paced introduction to some basic notions of commutative algebra and algebraic geometry. Many good books can be consulted for further details, but the text [Cox et al. 92] is certainly a good place to start. Our needs for tools of algebraic geometry are very modest. In particular, we only consider finite discrete affine varieties and the associated polynomials ring of functions. The combinatorial outlook on the relevant concepts can be traced back at least to Stanley's paper [Stanley 79], which is eminently readable. For more details also see his book [Stanley 96].

7.1 Ideals and Varieties

The context here is the ring $R = \mathbb{C}[\mathbf{x}]$ of complex coefficient polynomials in variables $\mathbf{x} = x_1, \ldots, x_n$. As usual, the ideal $\mathcal{I} = \langle f_1, \ldots, f_k \rangle$ of R is the set of expressions of the form $g_1 f_1 + \cdots + g_k f_k$ with $g_i \in R$. The set $V(\mathcal{I})$ of common solutions $\mathbf{a} \in \mathbb{C}^n$ to the equations $f_i(\mathbf{a}) = 0$ is called the *affine variety* defined by $\mathcal{I}$. Conversely, for a variety V we consider the ideal $I(V)$ of polynomials that vanish on all points of V, i.e., $f(\mathbf{a}) = 0$ for all $\mathbf{a} \in V$. It is clear that $\mathcal{I} \subseteq I(V)$ whenever $V = V(\mathcal{I})$. Thus, the ideal $I(V)$ is the set of all polynomials that are zero-valued on V, whereas the ideal $\mathcal{I}$ may very well contain only some of these. A simple example of this phenomenon occurs for the ideal $\mathcal{I} = \langle x^n \rangle$, with $n > 1$. Indeed $V(\mathcal{I}) = \{0\}$, so we get $I(V) = \langle x \rangle$, which is evidently larger than $\mathcal{I}$.

The ring $\mathbb{C}[V] := \mathbb{C}[\mathbf{x}]/I(V)$ is the ring of *polynomial functions* on V. This quotient is natural, since two polynomials define the same function on V if and only if their difference lies in $I(V)$. To make precise the notion of equivalence between sets of equations $\{g_j = 0\}_{1 \leq j \leq \ell}$, we are invariably led to state everything in terms of ideals. This is to say that a set of equations $\{g_j = 0\}_{1 \le j \le \ell}$ is equivalent to another set $\{f_i = 0\}_{1 \le i \le k}$ if and only if $\langle f_1, \ldots, f_k \rangle = \langle g_1, \ldots, g_\ell \rangle$. One of our purposes here is to describe special sets of generators for ideals, known as "Gröbner bases". Among the many nice properties of these special generator sets, one certainly stands out. This is the fact that the knowledge of a Gröbner basis for $I(V)$ makes possible an entirely explicit description of a linear basis for the ring $\mathbb{C}[V]$.

Fixing a monomial order (see Section 1.4), the *monomial ideal* $m(\mathcal{I})$ associated with $\mathcal{I}$ is the ideal generated by the leading monomials of elements of $\mathcal{I}$, i.e., $m(\mathcal{I}) = \langle m(f) \mid f \in \mathcal{I} \rangle$. For instance, with lexicographic order, the ideal $\langle x + y, x^2 + xy + y^2 \rangle$ has monomial ideal $\langle x, y^2 \rangle$. In general, for $\mathcal{I}$ generated by $f_1, \ldots, f_k$, we have an inclusion $\langle m(f_1), \ldots, m(f_k) \rangle \subseteq m(\mathcal{I})$, but this example illustrates that we can have strict inclusion, since the left-hand side is seen to be $\langle x, x^2 \rangle$.

Homogeneous Ideals

Essentially all the ideals that we are going to consider are *homogeneous*. This is to say that for $f \in \mathcal{I}$, we have $\pi_d(f) \in \mathcal{I}$, $0 \le d$. (See Section 3.1 for the definition of π_d.) For any ideal $\mathcal{I}$, let $\text{Gr}(\mathcal{I})$ be the homogeneous ideal generated by maximal degree homogeneous components of polynomials in $\mathcal{I}$, i.e., $\text{Gr}(\mathcal{I}) := \{\pi_{\deg(f)}(f) \mid f \in \mathcal{I}\}$. For $\mathcal{I}$ homogeneous we obviously have $m(\mathcal{I}) = m(\text{Gr}(\mathcal{I}))$. Let us denote by $\mathcal{B}_\mathcal{I}$ the set of monomials $\{\mathbf{x}^\mathbf{b} \mid \mathbf{x}^\mathbf{b} \notin m(\mathcal{I})\}$. Applying Gauss elimination (modulo $\mathcal{I}$) to the set of all monomials, we can check that $\mathcal{B}_\mathcal{I}$ is a set of representatives for a basis of $\mathbb{C}[\mathbf{x}]/\mathcal{I}$. Let us denote by $\mathcal{B}$ the basis obtained by this Gauss elimination process. By definition, if $\mathbf{x}^\mathbf{a}$ lies in $m(\mathcal{I})$ there is some polynomial $\mathbf{x}^\mathbf{a} + \sum_{\mathbf{b} \prec \mathbf{a}} c_\mathbf{b} \mathbf{x}^\mathbf{b}$ in $\mathcal{I}$ having $\mathbf{x}^\mathbf{a}$ as leading monomial. This forces $\mathbf{x}^\mathbf{a}$ to be eliminated in the the construction of $\mathcal{B}$, implying that $\mathcal{B} \subseteq \mathcal{B}_\mathcal{I}$. We deduce that $\mathcal{B}_\mathcal{I}$ spans $\mathbb{C}[\mathbf{x}]/\mathcal{I}$, and we need only check that it is linearly independent. For this, suppose that there is some linear combination $\sum_{\mathbf{x}^\mathbf{b} \in \mathcal{B}_\mathcal{I}} c_\mathbf{b} \mathbf{x}^\mathbf{b}$, sitting in $\mathcal{I}$, with nonzero coefficients. The leading monomial of this linear combination has to lie in $m(\mathcal{I}) \cap \mathcal{B}_\mathcal{I}$. This is empty by definition of $\mathcal{B}_\mathcal{I}$, thus we get a contradiction; hence $\mathcal{B}_\mathcal{I}$ is linearly independent. Observe that, for $\mathcal{I}$ homogeneous and $\mathcal{B}_\mathcal{I}$ finite, we have

$$\dim \mathbb{C}[\mathbf{x}]/\mathcal{I} = \dim \mathbb{C}[\mathbf{x}]/\text{Gr}(\mathcal{I}), \tag{7.1}$$

since $\mathcal{B}_\mathcal{I} = \mathcal{B}_{\text{Gr}(\mathcal{I})}$.

7.2   Gröbner Basis

From now on, we assume that we have chosen a fixed (graded) monomial order. A finite set $G = \{g_1, \ldots, g_\ell\}$ of monic polynomials is said to be a (reduced) *Gröbner basis* of the ideal $\mathcal{I}$, if we have

(1) $\mathcal{I} = \langle g_1, \ldots, g_\ell \rangle$,

(2) $m(\mathcal{I}) = \langle m(g_1), \ldots, m(g_\ell) \rangle$,

(3) no monomial of g_i is divisible by $m(g_j)$, for $j \neq i$.

There is a unique such basis, and it is explicitly computable from a finite generator set of $\mathcal{I}$. This calculation can be realized using Buchberger's algorithm, which has been implemented in most computer algebra systems.

Ideal Generated by Symmetric Polynomials

Let us illustrate the Gröbner basis construction with the ideal

$$\mathcal{I}_n := \langle h_1(\mathbf{x}), h_2(\mathbf{x}), \ldots, h_n(\mathbf{x}) \rangle$$

generated by the complete homogeneous symmetric polynomials. Denote by $\mathbf{x}_k$ the restricted set of variables $\mathbf{x}_k = \{x_k, x_{k+1}, \ldots, x_n\}$. We intend to show that the set $\{h_1(\mathbf{x}_1), h_2(\mathbf{x}_2), \ldots, h_n(\mathbf{x}_n)\}$ is a Gröbner basis for the ideal $\mathcal{I}_n$, under the lexicographic monomial order. For example, the set

$$h_1(\mathbf{x}_1) = x_1 + x_2 + x_3, \quad h_2(\mathbf{x}_2) = x_2^2 + x_2 x_3 + x_3^2, \quad h_3(\mathbf{x}_3) = x_3^3,$$

generates $\mathcal{I}_3$. We have already seen in Chapter 3 that the elementary symmetric polynomials e_k, $1 \leq k \leq n$, all lie in $\mathcal{I}_n$, so that

$$(1 - x_1 t) \cdots (1 - x_n t) \equiv 1 \mod \mathcal{I}_n.$$

This implies that for all $k \geq 1$, the identity

$$(1 - x_1 t) \cdots (1 - x_{k-1} t) = \frac{1}{(1 - x_k t) \cdots (1 - x_n t)} \tag{7.2}$$

holds in the ring $\mathcal{R}_n[\![t]\!]$, with $\mathcal{R}_n = \mathbb{C}[\mathbf{x}]/\mathcal{I}_n$. Comparing coefficients of t^k, we deduce that $h_k(\mathbf{x}_k) = 0$, since the kth power of t does not appear in the left-hand side. Recall that the expansion of $h_k(\mathbf{x}_k)$, as a polynomial in the variable x_k, is

$$h_k(\mathbf{x}_k) = x_k^k + \sum_{j=0}^{k-1} x_k^j h_{k-j}(\mathbf{x}_{k+1}). \tag{7.3}$$

Passing to the remainder modulo $h_k(\mathbf{x}_k)$, we get a reduction of x_k to a linear combination of powers of x_k in the range going from 0 to $k - 1$. It follows that the space $\mathcal{R}_n$ affords the set of monomials $x_1^{\epsilon_1} x_2^{\epsilon_2} \cdots x_n^{\epsilon_n}$, with $\epsilon_k < k$, as a generating set. We can now verify that $\{h_1(\mathbf{x}_1), h_2(\mathbf{x}_2), \ldots, h_n(\mathbf{x}_n)\}$ satisfies condition (3) of the definition of Gröbner basis. Observe that the corresponding set of leading monomials is $\{x_1, x_2^2, x_3^3, \ldots, x_n^n\}$. For $i < k$, x_k^k cannot divide terms of $h_i(\mathbf{x}_i)$, since they have smaller degrees. On the other hand, for $k > i$, $h_i(\mathbf{x}_i)$ does not contain the variable x_k, so none of its terms can be divisible by x_k^k. This shows that condition (3) holds.

Exercise. Show that the set $\{h_1(\mathbf{x}_1), h_2(\mathbf{x}_2), \ldots, h_n(\mathbf{x}_n)\}$ satisfies condition (2) of the definition of Gröbner basis for the ideal $\mathcal{I}_n$.

Using very similar arguments, a more general version of the statements above may be obtained:

Proposition 7.1 (Artin basis). *For any $k \geq 0$, the collection of monomials $x_1^{\epsilon_1} x_2^{\epsilon_2} \cdots x_n^{\epsilon_n}$, with $0 \leq \epsilon_i \leq k + i - 1$, constitutes a basis for the quotient ring $\mathbb{C}[\mathbf{x}]/J_{k,n}$, with*

$$J_{k,n} = \langle h_{k+1}, h_{k+2}, \cdots, h_{k+n} \rangle. \tag{7.4}$$

Indeed, this ideal affords the set $\{h_{k+1}(\mathbf{x}_1), h_{k+2}(\mathbf{x}_2), \ldots, h_{k+n}(\mathbf{x}_n)\}$ as a Gröbner basis.

7.3   Graded Algebras

Many of the spaces that we are going to study are naturally graded. Recall that a vector space (algebra) $\mathcal{V}$ (over a field $\mathbb{K}$) is said to be *graded* (or $\mathbb{N}$-*graded*) if we have a direct sum decomposition of $\mathcal{V}$ into a direct sum finite dimensional spaces:

$$\mathcal{V} = \mathcal{V}_0 \oplus \mathcal{V}_1 \oplus \mathcal{V}_2 \oplus \cdots, \tag{7.5}$$

with $\mathcal{V}_0 = \mathbb{K}$ and $\dim \mathcal{V}_i$ finite for all i. In the case of algebras, we also require that $\mathcal{V}_i \mathcal{V}_j \subseteq \mathcal{V}_{i+j}$, for all $0 \leq i, j$. Elements of $\mathcal{V}_d$ are called *homogeneous of degree d*, and the space $\mathcal{V}_d$ is the *degree d homogeneous component* of $\mathcal{V}$. The essence of equality (7.5) is that any element f of $\mathcal{V}$ affords a unique decomposition $f = f_0 + f_1 + f_2 + \cdots$ into its homogeneous components $f_d \in \mathcal{V}_d$. Every graded space $\mathcal{V}$ affords a *Hilbert series* $\text{Hilb}_t(\mathcal{V})$ of the $\mathcal{V}$, defined as $\sum_{d \geq 0} \dim(\mathcal{V}_d) t^d$.

Graded Representations

Even more interesting is the notion of *graded G-module*. Starting with a linear action of G on V, we assume that each $\mathcal{V}_d$ is G-invariant in (7.5). We get a graded decomposition of $\mathcal{V}$ into irreducible G-modules:

$$\mathcal{V} = \bigoplus_{d \geq 0} \bigoplus_{c \in \mathcal{C}} a_{c,d} \mathcal{V}^c,$$

with $a_{c,d}$ denoting the multiplicity of the irreducible G-module $\mathcal{V}^c$ in the degree d homogeneous component of $\mathcal{V}$. This leads us to introduce the notion of a *graded character* $\chi_{\mathcal{V},t} \colon G \longrightarrow \mathbb{C}[[t]]$ defined by

$$\chi_{\mathcal{V},t}(g) := \sum_{d \geq 0} \chi_{\mathcal{V}_d}(g) t^m. \tag{7.6}$$

This series in t contains all the information about the decomposition of each of the homogeneous components of $\mathcal{V}$ into irreducible G-modules. In other words, we can compute each of the multiplicities $a_{c,d}$ by taking the coefficient of t^d in the series resulting from the scalar product $\langle \chi_{\mathcal{V},t}(g), \chi_c \rangle$.

Graded Frobenius Characteristic

In the symmetric group case, we can further translate in terms of Frobenius characteristics. More precisely, the *graded Frobenius characteristic* $\mathcal{F}_t(\mathcal{V})$ is the formal power series $\sum_{d \geq 0} \mathcal{F}(\mathcal{V}_d) t^d$, with coefficient in the ring $\mathbb{C}[\mathbf{x}]^{\mathfrak{S}}$ of symmetric functions in the variables $\mathbf{z} = z_1, z_2, \ldots$. This is evidently a polynomial in t when $\mathcal{V}$ is of finite dimension. Expanding each coefficient in terms of Schur functions, we get $\mathcal{F}_t(\mathcal{V}) = \sum_\lambda a_\lambda(t) s_\lambda$, where each coefficient $a_\lambda(t) = \sum_{d \geq 0} a_{\lambda,d} t^d$ is a positive integer coefficient series in t, with $a_{\lambda,d}$ giving the multiplicity of the irreducible representation associated with s_λ in the homogeneous component $\mathcal{V}_d$. Let us illustrate with the graded $\mathfrak{S}_n$-module $R = \mathbb{C}[\mathbf{x}]$, of polynomials in n variables $\mathbf{x} = x_1, x_2, \ldots, x_n$. The associated graded Frobenius characteristic is

$$\mathcal{F}_t(R) = h_n \left[\frac{\mathbf{z}}{1 - t} \right]. \tag{7.7}$$

This is shown as follows. We first recall that by definition

$$\mathcal{F}_t(R) = \sum_{d \geq 0} t^d \frac{1}{n!} \sum_{\sigma \in \mathfrak{S}_n} \chi_{R_d}(\sigma) p_{\lambda(\sigma)}(\mathbf{z}).$$

Let us compute $\chi_{R_d}(\sigma)$ in the basis of monomials. This corresponds to counting degree d monomials $\mathbf{x}^{\mathbf{a}}$ (i.e., $|\mathbf{a}| = d$) that are fixed by σ, i.e., $\sigma \cdot \mathbf{x}^{\mathbf{a}} = \mathbf{x}^{\mathbf{a}}$. This forces $\mathbf{a} = (a_1, \ldots, a_n)$ to be such that $a_j = k_\gamma$ for all j appearing in a same cycle γ of σ, with $k_\gamma \in \mathbb{N}$ depending only on γ. If the cycle type of σ is $\lambda = \lambda(\sigma)$, it follows that the required enumeration is globally obtained in the form

$$\sum_d \chi_{R_d}(\sigma) t^d = \prod_i \frac{1}{1 - t^{\lambda_i}},$$

using the variable t to keep track of degree. Observing that the right-hand side of this identity is just $p_\lambda[1/(1-t)]$, we get equation (7.7).

7.4   The Cohen–Macaulay Property

Most of the rings and modules that we consider in our context have very special structures, implying that their Hilbert series has a rather special form. Their common property is that they are "Cohen–Macaulay" as described below. For a more general and full description of the notions considered here we refer to the classic textbook [Atiyah and Macdonald 69].

Homogeneous System of Parameters

The *Krull dimension* of a graded module $\mathcal{M}$ is the maximum number of homogeneous elements of $\mathcal{M}$ that are algebraically independent over the ground field. In the case of Krull dimension r, a sequence $\theta_1, \theta_2, \ldots, \theta_r$ is said to be a *homogeneous system of parameters* (HSOP) if $\mathcal{M}$ is a finitely-generated module over the ring of polynomials in the θ_i. For example, the ring $\mathbb{C}[\mathbf{x}]^{\mathfrak{S}_n}$ of symmetric polynomials is of Krull dimension n, and it affords as HSOP (among others) any of the three sets $\{e_1, \ldots, e_n\}$, $\{h_1, \ldots, h_n\}$, or $\{p_1, \ldots, p_n\}$. Similar statements holds for any reflection group G.

Cohen–Macaulay

In some instance the module $\mathcal{M}$ is very well behaved with respect to HSOP, in that there is a simple decomposition of $\mathcal{M}$ of the form

$$\mathcal{M} \simeq \mathcal{M}/\mathcal{I} \otimes \mathbb{C}[\theta_1, \ldots, \theta_r], \tag{7.8}$$

with $\theta_1, \ldots, \theta_r$ an HSOP for $\mathcal{M}$, and $\mathcal{I}$ the ideal generated by this HSOP. We also require that the isomorphism in (7.8) be a graded module isomorphism. When we have (7.8) holding, there exists a family $\varphi_1, \ldots, \varphi_m$ of elements of $\mathcal{M}$ such that any element f of $\mathcal{M}$ can be expressed in a unique manner in the form $f = a_1 \varphi_1 + \cdots + a_m \varphi_m$, with the coefficients a_j belonging to the ring of polynomials in the θ_i. When this is the case, we say that $\mathcal{M}$ is *Cohen–Macaulay*. It immediately follows from this decomposition that the Hilbert series of $\mathcal{M}$ has the rather simple form

$$\mathrm{Hilb}_t(\mathcal{M}) = \frac{\alpha(t)}{(1 - t^{d_1}) \cdots (1 - t^{d_r})}, \tag{7.9}$$

with the d_i equal to the degree of θ_i and $\alpha(t)$ equal to the Hilbert series (which is, in fact, a polynomial) of $\mathcal{M}/\mathcal{I}$.

Chapter 8

Coinvariant Spaces

We will now explore one of the central motivating results for this book. The set of classical results of representation theory discussed in this chapter will be the inspiration for much of the remaining chapters. At the heart of these results sits the notion of covariant space of a group of (orthogonal) $n \times n$ matrices G. One possible explanation for the role of this space stems from the study of the G-module decomposition of the space $\mathbb{R}[\mathbf{x}]$ of polynomials in n variables. The story begins with the observation that we can trivially get many copies of any known irreducible component $\mathcal{V}$ of $\mathbb{R}[\mathbf{x}]$, simply by tensoring $\mathcal{V}$ with a given G-invariant polynomial. This naturally raises the question of finding a basic set of irreducible components out of which all others can be obtained through this means. This is precisely what the G-covariant space will give us in one stroke. For more on this fascinating subject see [Kane 01], or [Garsia and Haiman 08].

8.1   Coinvariant Spaces

Our previous introductory comment makes it natural to mod out the space of polynomials by the ideal generated by G-invariant polynomials, with G a finite group of $n \times n$ orthogonal matrices. Thus we obtain an important companion to the ring of G-invariants, namely R_G, the G-*coinvariant space*, as the quotient $\mathbb{R}[\mathbf{x}]/\mathcal{I}_G$, where $\mathcal{I}_G$ denotes the ideal of $R = \mathbb{R}[\mathbf{x}]$ generated by constant-term free elements of R^G. Observing that $\mathcal{I}_G$ is a homogeneous subspace of R, we deduce that the ring R_G is naturally graded by degree. Moreover, the group G acts naturally on R_G. In fact, it can be shown that the G-module R_G is actually isomorphic to the *left regular representation* of G if and only if G is a group that is generated by its reflections.[1]   In particular, the dimension of R_G is exactly the order of the group G. The

---

[1]See [Kane 01] or [Stanley 79].

link between R^G and R_G is established through Chevalley's theorem, which states that there is a natural isomorphism of graded G-modules

$$R \simeq R^G \otimes R_G. \tag{8.1}$$

We will give a general outline of a proof in this chapter. One can decode (8.1) as giving a precise formulation for the somewhat vague statement:

> *Any irreducible component of R can be obtained by tensoring one irreducible component of R_G by some G-invariant polynomial.*

One immediate consequence of this, in view of (3.4) and (3.10), is that

$$\mathrm{Hilb}_t(R_G) = \prod_{i=1}^{n} \frac{1 - t^{d_i}}{1 - t}$$
$$= (1 + \cdots + t^{d_1 - 1}) \cdots (1 + \cdots + t^{d_n - 1}), \tag{8.2}$$

with the d_i corresponding to the respective degrees of any *basic set* of G-invariants $\{f_1, \ldots, f_n\}$. Recall that this means $\mathcal{I}_G = \langle f_1, \ldots, f_n \rangle$ with the f_i algebraically independent. It is interesting that the spaces R_G also arise in a theorem of Borel as the cohomology rings of flag manifolds. We derive from (8.1) that

$$\chi_{\mathcal{R}}(t) = \mathrm{Hilb}_t(R^G) \chi_{\mathcal{R}_G}(t)$$
$$= \prod_{i=1}^{n} \frac{1}{1 - t^{d_i}} \chi_{\mathcal{R}_G}(t). \tag{8.3}$$

Observe that in the case of $G = \mathfrak{S}_n$, the dimension of R_G is $n!$. In a sense this is the $n!$ that serves as a leitmotiv for this book. The degrees d_i in this case are the integers from 1 to n, since we can choose the symmetric polynomials $e_1, \ldots, e_n$ as a basic set of $\mathfrak{S}_n$-invariants. The Hilbert series of the space $R_n := R_{\mathfrak{S}_n}$ is thus the t-analog of $n!$ (see Section 1.1). Proposition 7.1 gives a description of a linear basis of representatives for the quotient R_n. For $n = 3$, these are the six monomials $1, x_2, x_3, x_2 x_3, x_3^2, x_2 x_3^2$. In this case, degree enumeration does give $1 + 2t + 2t^2 + t^3 = (1 + t)(1 + t + t^2)$, as announced.

An Explicit Example

Before going on with the general setup, let us work out the case $n = 2$, choosing the group to be $\mathfrak{S}_2$ (there are not a lot of choices). With $R = \mathbb{R}[x, y]$, the ring $R^{\mathfrak{S}_2}$ is generated by the two algebraically independent polynomials $x + y$ and xy, so that all symmetric polynomials are linear combinations of $e_1^k e_2^\ell = (x+y)^k (xy)^\ell$. Modulo the ideal $\mathcal{I}_2$ generated by e_1 and e_2, it is evident that $x^k y^\ell \equiv (-1)^\ell x^{k+\ell}$, since $x + y \equiv 0 \mod \mathcal{I}_2$. Taking $k = \ell = 1$ or $k = 0$ and $\ell = 2$, we may check that all degree 2 monomials are in the ideal $\mathcal{I}_2$, hence all monomials of degree ≥ 2 are in the ideal. The quotient R_G can thus be identified with the linear span of $\{1, y\}$, and we have thus naively rechecked Proposition 7.1.

If we are to believe (8.1), then any two-variable polynomial $f(x, y)$ (say homogeneous, for simplicity) can be expanded in a unique fashion as a linear combination of polynomials either of the form $e_1^k e_2^\ell$ or $y e_1^{k-1} e_2$. Considering the degree d homogeneous components R_d of R, this says that we have the following linear basis:

(1) the polynomials $y e_1$, e_1^2 and e_2 span the dimension 3 component R_2,

(2) the polynomials $y e_1^2$, e_1^3, $y e_2$ and $e_1 e_2$ span the dimension 4 component R_3,

(3) the polynomials $y e_1^3$, e_1^4, $y e_1 e_2$, $e_1^2 e_2$ and e_2^2 span the dimension 5 component R_3, etc.

Exercise. Find a formula for the expansion of degree d homogeneous two-variable polynomials in terms of the linear basis just described.

8.2 Harmonic Polynomials

The coinvariant space of G is closely related to another important G-module, the space $\mathcal{H}_G$ of G-*harmonic polynomials*, which is defined as the orthogonal complement $\mathcal{I}_G^\perp$ of the ideal $\mathcal{I}_G$. Unfolding this definition, with the scalar product as defined in (3.5), $g(\mathbf{x})$ is G-harmonic if and only if it satisfies the system of n partial differential equations $f_k(\partial \mathbf{x}) p(\mathbf{x}) = 0$, one for each of the f_k in a basic set of G-invariants. Recall that the scalar product is G-invariant, thus we have a natural action of G on $\mathcal{H}_G$. The central fact that we will exploit is that the spaces $\mathcal{H}_G$ and R_G are actually isomorphic as graded G-modules [Steinberg 64].

As it happens, there is a very nice explicit description of the space $\mathcal{H}_G$ using the fact that it is closed under partial derivatives and that it contains the Jacobian determinant $\Delta_G(\mathbf{x})$ (see Section 3.4, definition (3.11)). To check this last assertion, we observe that the polynomial $f(\partial \mathbf{x}) \Delta_G(\mathbf{x})$ has to be G-skew-invariant, for any constant term free G-invariant polynomial $f(\mathbf{x})$. Thus, it must be divisible by the Jacobian determinant polynomial $\Delta_G(\mathbf{x})$ (see Section 3.4). However, its degree is strictly smaller than that of $\Delta_G(\mathbf{x})$, which forces $f(\partial \mathbf{x}) \Delta_G(\mathbf{x}) = 0$. We conclude that Δ_G is a G-

harmonic polynomial. In turn, all polynomials $\partial \mathbf{x}^{\mathbf{a}} \Delta_G(\mathbf{x})$ are also necessarily G-harmonic. A little more work shows that we have obtained in this way a complete and simple description of the space of G-harmonic polynomials. More precisely, we will see that there can be no G-harmonic polynomials other than those that are obtained as linear combinations of polynomials of the form $\partial \mathbf{x}^{\mathbf{a}} \Delta_G(\mathbf{x})$, essentially by showing that $\dim \mathcal{H}_G < |G|$.

To sum up, we can formulate our global assertion as $\mathcal{H}_G = \mathcal{L}_\partial[\Delta_G(\mathbf{x})]$, using $\mathcal{L}_\partial$ as shorthand for "linear span of all partial derivatives of". We can also reformulate the decomposition given in (8.1) as $R = R^G \otimes \mathcal{H}_G$. The point of the equality here is that both R^G and $\mathcal{H}_G$ being G-submodules of R, we can have actual equality, meaning that there is a canonical decomposition of any polynomial $p(\mathbf{x})$ in the form $p(\mathbf{x}) = \sum_{\gamma \in G} f_\gamma(\mathbf{x}) \partial \mathbf{x}^{\mathbf{a}(\gamma)} \Delta_G(\mathbf{x})$. Here the $\mathbf{a}(\gamma)$ are chosen so that the set of $\partial \mathbf{x}^{\mathbf{a}(\gamma)} \Delta_G(\mathbf{x})$ forms a basis of $\mathcal{H}_g$, and the f_γ are G-invariant polynomials. We will see in particular instances that there are many interesting choices for the vectors $\mathbf{a}(\gamma)$.

8.3   Regular Point Orbits

To get an $n!$ upper bound for the dimension of $\mathcal{H}_{\mathfrak{S}_n}$, we exploit its natural link with $R_n = R_{\mathfrak{S}_n}$, the coinvariant space for $\mathfrak{S}_n$. In Proposition 7.1, we gave a spanning set of $n!$ monomials for the space R_n. In other words, $\dim R_n \leq n!$, the evident vector space isomorphism between $\mathcal{I}_n^\perp$ and $R/\mathcal{I}_n$, gives $\dim \mathcal{H}_n \leq n!$. Similar arguments show that we have the same kind of inequality for all reflection groups G. To show that we actually have equality, we need to deepen our understanding of the covariant space of G. This is elegantly achieved using basic tools of algebraic geometry. So consider the *orbit* $G\mathbf{v}$ of a point $\mathbf{v} = (v_1, v_2, \ldots, v_n) \in \mathbb{R}^n$ under the action of G, i.e., $G\mathbf{v} := \{\gamma \cdot \mathbf{v} \mid \gamma \in G\}$. The point (or the associated orbit) is said to be *regular* if it has $|G|$ elements. For instance, with $G = \mathfrak{S}_3$, we could choose $\mathbf{v} = (1, 2, 3)$, and then

$$G\mathbf{v} = \{(1,2,3), (2,1,3), (2,3,1), (3,2,1), (3,1,2), (1,3,2)\}.$$

Now we consider the set $G\mathbf{v}$ as a (discrete) algebraic variety, and the corresponding ring $R_{G\mathbf{v}} = R/\mathcal{I}_{G\mathbf{v}}$ of polynomial functions on $G\mathbf{v}$. We are going to suppose, from now on, that $\mathbf{v}$ is a regular point. In particular, this makes it apparent that $|G|$ is the dimension of $R_{G\mathbf{v}}$, since it is the space of functions on the $|G|$-element set $G\mathbf{v}$. We may best understand this by exhibiting a basis of $R_{G\mathbf{v}}$ as follows. There is a unique (modulo $\mathcal{I}_\mathbf{v}$) polynomial $\psi(\mathbf{x})$, of minimal degree that takes the value 1 at point $\mathbf{v}$ and 0 at all other points of $G\mathbf{v}$. In the case $G = \mathfrak{S}_n$, this is just the *Lagrange interpolation polynomial* $\prod_{i<j} (x_i - v_j)/(v_i - v_j)$. For $\gamma \in G$, we can then

set $\varphi_\gamma(\mathbf{x}) := \psi(\gamma \cdot \mathbf{x})$, thus obtaining a G indexed family of polynomials, such that

$$\varphi_\gamma(\mathbf{u}) = \begin{cases} 1 & \text{if } \mathbf{u} = \gamma^{-1} \cdot \mathbf{v}, \\ 0 & \text{otherwise.} \end{cases}$$

The set $\{\varphi_\tau\}_{\gamma \in G}$ clearly forms a linear basis for $R_{G\mathbf{v}}$, so that $\dim R_{G\mathbf{v}} = |G|$. Moreover, $\gamma \cdot \varphi_\tau(\mathbf{x}) = \varphi_{\gamma\tau}(\mathbf{x})$, showing explicitly that G acts on $R_{G\mathbf{v}}$ just as G acts on $\mathbb{R}G$ by left multiplication. This makes it evident that $R_{G\mathbf{v}}$ is isomorphic to the regular representation of G.

Orbit Harmonics

To finish the characterization of $\mathcal{H}_G$, we consider the space of *orbit harmonics*, $\mathcal{H}_{G\mathbf{v}}$, defined for a regular point $\mathbf{v}$ as $\mathcal{H}_{G\mathbf{v}} := \mathrm{Gr}(\mathcal{I}_{G\mathbf{v}})^\perp$. Observe that $\mathcal{I}_G \subseteq \mathrm{Gr}(\mathcal{I}_{G\mathbf{v}})$, since, for any f homogeneous and G-invariant, the polynomial $f(\mathbf{x}) - f(\mathbf{v})$ vanishes at all points of $G\mathbf{v}$. Moreover, the maximal degree homogeneous component of $f(\mathbf{x}) - f(\mathbf{v})$ is clearly $f(\mathbf{x})$, and thus $f(\mathbf{x})$ lies in $\mathrm{Gr}(\mathcal{I}_{G\mathbf{v}})$. It follows immediately that $\mathcal{H}_{G\mathbf{v}} \subseteq \mathcal{H}_G$. Thus, $\dim \mathcal{H}_G \geq |G|$. In fact, according to a theorem of R. Steinberg [Steinberg 64], there is equality if and only if G is a group generated by reflections. More details can be found in [Kane 01].

8.4   Symmetric Group Harmonics

Let us discuss the notion of G-harmonic polynomials in further detail in the case where G is the symmetric group $\mathfrak{S}_n$. As already mentioned, we simplify the notation to $\mathcal{H}_n$ in this context, and simply say *harmonic polynomials*, rather than $\mathfrak{S}_n$-harmonic polynomials. As a word of explanation for the terminology, recall that the classical notion of harmonic function is that of solutions of $\partial x_1^2 f(\mathbf{x}) + \cdots + \partial x_n^2 f(\mathbf{x}) = 0$. On the other hand, for a polynomial to be $\mathfrak{S}_n$-harmonic we need it to satisfy the set of partial differential equations

$$(\partial x_1 + \cdots + \partial x_n)f(\mathbf{x}) = 0,$$
$$(\partial x_1^2 + \cdots + \partial x_n^2)f(\mathbf{x}) = 0,$$
$$\vdots$$
$$(\partial x_1^k + \cdots + \partial x_n^k)f(\mathbf{x}) = 0,$$

with k at least equal to the degree of $f(\mathbf{x})$. Clearly, we get a stronger notion of harmonicity than the classical one. Let us illustrate by giving an explicit basis for the space $\mathcal{H}_3$ of $\mathfrak{S}_3$-harmonics:

Basis elements	Graded Frobenius characteristic
$(x_1 - x_2)(x_1 - x_3)(x_2 - x_3)$	$t^3 s_{111}$
$(x_1 - x_2)(x_1 + x_2 - 2x_3), (x_1 - x_3)(x_1 + x_3 - 2x_2)$	$t^2 s_{21}$
$(x_1 - x_2), (x_1 - x_3)$	$t s_{21}$
1	$s_3.$

It follows that the graded Frobenius characteristic of $\mathcal{H}_3$ is $\mathrm{Frob}_t(\mathcal{H}_3) = s_3 + (t + t^2) s_{21} + t^3 s_{111}$. In general, we get a basis of $\mathcal{H}_n$ by considering the $n!$ polynomials $\partial \mathbf{x}^\epsilon \Delta_n(\mathbf{x})$, corresponding to choosing the $n!$ exponent vectors $\epsilon = (\epsilon_1, \ldots, \epsilon_n)$ such that $0 \leq \epsilon_k \leq n - k$. The resulting leading monomials are clearly $x_1^{n-1-\epsilon_1} x_2^{n-2-\epsilon_2} \cdots x_n^{n-n-\epsilon_n}$. Since they are all distinct, we do get a basis of $\mathcal{H}_n$ in view of the discussion of the previous sections.

An interesting feature of the space $\mathcal{H}_n$ is that it contains each of the irreducible $\mathfrak{S}_n$-modules $\mathcal{S}^\lambda$ constructed in Section 5.5. To check this, we exhibit a monomial differential operator $\partial \mathbf{x}^{\mathbf{a}(\lambda)}$ such that $\partial \mathbf{x}^{\mathbf{a}(\lambda)} \Delta_n(\mathbf{x}) = c \Delta_\tau(\mathbf{x})$ for some nonzero constant c, and some standard tableau τ of shape λ. The exponent vector $\mathbf{a}(\tau) = (a_1, \ldots, a_n)$ is obtained by choosing a_i, the exponent of x_i, to be equal to $i - 1 - \eta(i)$, with $\eta(i)$ being the height of i in τ. For example, taking λ to be the shape

we check that

$$\partial x_3^2 \partial x_4^2 \partial x_5^4 \Delta_5(\mathbf{x}) = 288(x_1 - x_1)(x_4 - x_3).$$

As we have already announced, and will check in the next section, $\mathcal{H}_n$ is isomorphic to the regular representation. Hence, f^λ is the actual multiplicity of $\mathcal{S}^\lambda$ in $\mathcal{H}_n$. Isomorphic copies of $\mathcal{S}^\lambda$ appear in various homogeneous components, and we have exhibited the unique copy that appears in the lowest degree homogeneous component.

8.5 Graded Frobenius Characteristic of the $\mathfrak{S}_n$-Coinvariant Space

Another direct consequence of our discussion surrounding the graded $\mathfrak{S}_n$-module isomorphism $R \simeq R^{\mathfrak{S}_n} \otimes R_{\mathfrak{S}_n}$, is an explicit formula for the graded

Frobenius characteristic of the $\mathfrak{S}_n$-covariant space, as well as that of the space $\mathcal{H}_n$ of $\mathfrak{S}_n$-harmonics. This is

$$\text{Frob}_t(R_{\mathfrak{S}_n}) = \text{Frob}_t(\mathcal{H}_n) = \prod_{j=1}^{n}(1 - t^j)h_n\left[\frac{\mathbf{z}}{1 - t}\right], \qquad (8.4)$$

since we already know that $\text{Frob}_t(R) = h_n[\mathbf{x}/(1 - t)]$, and that the graded enumeration of $R^{\mathfrak{S}_n}$ is given by $\prod_{i=1}^{n}(1 - t^i)^{-1}$. There are two easy consequences of this. First, we expand the right-hand side, substituting $\mathbf{z}$ for $\mathbf{x}$ and $1/(1 - t)$ for $\mathbf{y}$ in the formula $h_n(\mathbf{xy}) = \sum_{\mu} p_\mu(\mathbf{x})p_\mu(\mathbf{y})/z_\mu$, to get

$$\prod_{j=1}^{n}(1 - t^j)h_n\left[\frac{\mathbf{z}}{1 - t}\right] = \prod_{j=1}^{n}(1 - t^j)\sum_{\mu \vdash n}\frac{1}{z_\mu}\prod_{i=1}^{\ell(\mu)}\frac{p_{\mu_i}(\mathbf{z})}{1 - t_i^\mu}. \qquad (8.5)$$

It then becomes manifest that the coefficient of $p_1(\mathbf{z})^n/n!$ in the right-hand side of this last identity is nothing else than $\prod_{j=1}^{n}(1 - t^j)/(1 - t)$, which is precisely the graded Hilbert series that we announced. Almost as clear is the fact that the limit, as q goes to 1, of the right-hand side of (8.5) must be $h_1^n = p_1^n$. Indeed, all terms but the one that corresponds to the partition 1^n vanish, since a factor of $(1 - t)^{n-\ell(\mu)}$ comes out of the product $\prod_{j=1}^{n}(1 - t^j)\prod_{i=1}^{\ell(\mu)}(1 - t_i^\mu)^{-1}$, and $\ell(\mu) < n$ except when $\mu = 1^n$. The $n!$ cancels out with the values of the limit in the case $\mu = 1^n$. The gist of this is that we get the character of the regular representation by the specialization $t = 1$, so that indeed both spaces $R_{\mathfrak{S}_n}$ and $\mathcal{H}_n$ are isomorphic to the regular representation.

8.6 Generalization to Line Diagrams

Given a strictly decreasing sequence of nonnegative integers $\mathbf{a}$ $(a_1 > a_2 > \cdots > a_n \geq 0)$, we have already considered in Section 3.9 the determinant $\Delta_{\mathbf{a}}(\mathbf{x}) = \det(x_i^{a_j})_{i,j=1}^{n}$. Recall that we may always write $\mathbf{a}$ in the form $\mathbf{a} = \lambda + \delta_n$ with λ a partition, and that the Schur polynomials where originally defined to be such that $\Delta_{\lambda+\delta_n}(\mathbf{x})$ factors as $s_\lambda(\mathbf{x})\Delta_n(\mathbf{x})$, where $\Delta_n = \Delta_{\delta_n}$ is the classical Vandermonde determinant. Now, denote by $\mathcal{H}_{\mathbf{a}} := \mathcal{L}_\partial(\Delta_{\mathbf{a}})$ the vector space spanned by all partial derivatives of $\Delta_{\mathbf{a}}$ (of any order). Our goal here is to show that $\mathcal{H}_{\mathbf{a}}$ carries a multiple of the left regular representation of $\mathfrak{S}_n$ and obtain an explicit expression for its graded Frobenius characteristic. Observe that $\mathcal{H}_n = \mathcal{L}_\partial(\Delta_n)$ is the graded version of the left regular representation, whose graded Frobenius characteristic is given in equation (8.4). Let Σ_λ denote the graded vector space spanned by the skew Schur functions $s_{\lambda/\mu}(\mathbf{x})$ as μ varies in the set of partitions included in λ. This space is also graded with respect to degree. We have the following explicit description of how the spaces $\mathcal{H}_{\lambda+\delta_n}$ decompose as graded $\mathfrak{S}_n$-modules. This makes more explicit part of a theorem of Stanley [Stanley 79, Proposition 4.9] that is rather stated in the context of general reflection groups. When not mentioned explicitly, all results of this section are joint with A. Garsia and G. Tesler; see [Bergeron et al. 00], where missing proofs can be found.

Theorem 8.1. *The graded Frobenius characteristic of the space $\mathcal{H}_{\lambda+\delta_n}$ is* $\mathrm{Hilb}_t(\Sigma_\lambda)\,\mathrm{Frob}_t(\mathcal{H}_n)$. *In particular, $\mathcal{H}_{\lambda+\delta_n}$ carries a multiple of the left regular representation that is equal to the dimension of Σ_λ.*

Before our discussion of why this theorem holds, let us go over some auxiliary facts and observations. First, for $\mathbf{b}$ $(= b_1 > b_2 > \cdots > b_n \geq 0)$, we have the following.

Lemma 8.2. *The polynomial $\Delta_\mathbf{b}(\partial\mathbf{x})\Delta_\mathbf{a}(\mathbf{x})$ is different from 0 only if $b_1 \leq a_1, \ldots, b_n \leq a_n$. In particular, the polynomials $\Delta_\mathbf{a}(\mathbf{x})$ constitute an orthogonal basis of the set of antisymmetric polynomials of $R = \mathbb{R}[\mathbf{x}]$.*

Proof: To see this, we first observe that $\Delta_\mathbf{b}(\partial\mathbf{x})\Delta_\mathbf{a}(\mathbf{x})$ expands as

$$\sum_{\sigma,\tau\in\mathfrak{S}_n} \varepsilon(\tau)\sigma\partial\mathbf{x}^{\tau\mathbf{b}}\mathbf{x}^\mathbf{a}.$$

Thus from $\Delta_\mathbf{b}(\partial\mathbf{x})\Delta_\mathbf{a}(\mathbf{x}) \neq 0$ we deduce that there must be at least one permutation τ such that $b_{\tau_i} \leq a_i$, for all $1 \leq i \leq n$. If this happens, a fortiori we must have the inequalities $b_i \leq a_i$. As for the last assertion, note that if $\Delta_\mathbf{a}$ and $\Delta_\mathbf{b}$ have different degrees then the orthogonality is trivial. On the other hand, if they have the same degree, then the nonvanishing of the scalar product implies that $\Delta_\mathbf{b}(\partial\mathbf{x})\Delta_\mathbf{a}(\mathbf{x}) \neq 0$, and also the inequalities $b_i \leq a_i$, which in this case forces $\mathbf{a} = \mathbf{b}$. $\square$

Lemma 8.3. *If $f(\mathbf{x})$ is a homogeneous symmetric polynomial, then*

$$f(\partial\mathbf{x})\Delta_n(\partial\mathbf{x})\Delta_\mathbf{a}(\mathbf{x}) \neq 0 \text{ if and only if } f(\partial\mathbf{x})\Delta_\mathbf{a}(\mathbf{x}) \neq 0.$$

When this is the case, we can find a homogeneous symmetric polynomial $g(\mathbf{x})$ giving $g(\partial\mathbf{x})f(\partial\mathbf{x})\Delta_\mathbf{a}(\mathbf{x}) = c\Delta_n(\mathbf{x})$, with $c \neq 0$.

Proof: The homogeneous polynomial $\varphi(\mathbf{x}) := f(\partial\mathbf{x})\Delta_\mathbf{a}(\mathbf{x})$ is clearly alternating, hence it factors as $\varphi(\mathbf{x}) = \psi(\mathbf{x})\Delta_n(\mathbf{x})$, with ψ symmetric. When $\varphi \neq 0$, the expression $\varphi(\partial\mathbf{x})\varphi(\mathbf{x})$ must not vanish since it evaluates to the sum of the square of the coefficients of φ. We thus have a nonzero polynomial $\varphi(\partial\mathbf{x})\varphi(\mathbf{x}) = \psi(\partial\mathbf{x})\Delta_n(\partial\mathbf{x})f(\partial\mathbf{x})\Delta_\mathbf{a}(\mathbf{x})$, which we may write as $\psi(\partial\mathbf{x})h(\mathbf{x})$, where $h(\mathbf{x})$ denotes the (clearly symmetric) polynomial $\Delta_n(\partial\mathbf{x})f(\partial\mathbf{x})\Delta_\mathbf{a}(\mathbf{x})$. Using the same reasoning as above, we see that we must also have $h(\partial\mathbf{x})h(\mathbf{x}) \neq 0$, thus $h(\partial\mathbf{x})\Delta_n(\partial\mathbf{x})f(\partial\mathbf{x})\Delta_\mathbf{a}(\mathbf{x})$ cannot vanish. In particular, we get that $h(\partial\mathbf{x})f(\partial\mathbf{x})\Delta_\mathbf{a}(\mathbf{x})$ is a nonzero alternating polynomial of degree

$$\deg\big(h(\partial\mathbf{x})f(\partial\mathbf{x})\Delta_\mathbf{a}(\mathbf{x})\big) = |\mathbf{a}| - \deg(f) - \left(|\mathbf{a}| - \deg(f) - \binom{n}{2}\right) = \binom{n}{2},$$

which forces $h(\partial\mathbf{x})f(\partial\mathbf{x})\Delta_\mathbf{a}(\mathbf{x})$ to be a nonzero multiple of the Vandermonde determinant $\Delta_n(\mathbf{x})$. This shows that we may choose $g(\mathbf{x}) = h(\mathbf{x})$, and our proof is complete. □

We are now in a position to deal with the special case of decreasing sequences of the form $\mathbf{a} = \mathbf{k}^n + \delta_n$, such as

$$2^4 + \delta_4 = (2,2,2,2) + (3,2,1,0) = (5,4,3,2),$$

which is both interesting in its own right and useful in our further developments. To begin with, note that in any case the orthogonal complement of our space $\mathcal{H}_\mathbf{a}$ is the ideal $I_\mathbf{a}$ of polynomial differential operators that kill $\Delta_\mathbf{a}$. In symbols, $I_\mathbf{a} = \big(f(\mathbf{x}) \mid f(\partial\mathbf{x})\Delta_\mathbf{a}(\mathbf{x}) = 0\big)$. In particular, we also have $I_\mathbf{a}^\perp = \mathcal{H}_\mathbf{a}$. Now we have the following.

Proposition 8.4. *The ideal $I_{\mathbf{k}^n+\delta_n}$ affords the polynomials h_{k+i}, for $1 \leq i \leq n$, as generators.*

Proof: For convenience, let us set $\mathbf{a} := \mathbf{k}^n + \delta_n$. Note that $h_{k+i}(\mathbf{x})\Delta_n(\mathbf{x}) = \Delta_{(k+i)+\delta_n}(\mathbf{x})$, since $h_{k+i} = s_{(k+i)}$. We apply Lemma 8.2 and conclude that $h_{k+i}(\partial\mathbf{x})\Delta_n(\partial\mathbf{x})\Delta_\mathbf{a}(\mathbf{x})$ must necessarily vanish for all $i \geq 1$. In turn, Lemma 8.3 gives that $h_{k+i}(\partial\mathbf{x})\Delta_\mathbf{a}(\mathbf{x}) = 0$, so we must have $h_{k+i}(\mathbf{x}) \in I_\mathbf{a}$ for all $i \geq 1$. In particular, we deduce the inclusion of ideals

$$J_{k,n} = (h_{k+1}, h_{k+2}, \ldots, h_{k+n}) \subseteq J_k = (h_{k+1}, h_{k+2}, \ldots) \subseteq I_\mathbf{a}. \tag{8.6}$$

Combining Proposition 7.1 with the inclusions in (8.6), we are led to the string of inequalities

$$\dim \mathbb{Q}[\mathbf{x}]/I_\mathbf{a} \leq \dim \mathbb{Q}[\mathbf{x}]/J_k \leq \dim \mathbb{Q}[\mathbf{x}]/J_{k,n} \leq k+n)!/k!. \tag{8.7}$$

On the other hand, the lexicographically leading term of $\Delta_\mathbf{a}$ is the monomial $\mathbf{x}^\mathbf{a}$. Thus, we see that differentiating $\Delta_\mathbf{a}$ by all the submonomials of its leading term will yield $(k+n)!/k!$ independent elements of $\mathcal{H}_\mathbf{a}$. So we must also have the inequality $(k+n)!/k! \leq \dim \mathcal{H}_\mathbf{a}$. But from $I_\mathbf{a}^\perp = \mathcal{H}_\mathbf{a}$ we deduce that $\dim \mathbb{Q}[\mathbf{x}]/I_\mathbf{a} = \dim \mathcal{H}_\mathbf{a}$. Combined with (8.7), we conclude that all these inequalities must be equalities, forcing the inclusions in (8.6) to be equalities as well, proving the proposition. □

Note that as a byproduct of our argument we get the following remarkable fact. In the case $\mathbf{a} = \mathbf{k}^n + \delta_n$, a basis $\mathcal{B}_{\mathbf{a}}$ for the space $\mathcal{H}_{\mathbf{a}}$ is given by the polynomials $\partial\mathbf{x}^\epsilon \Delta_{\mathbf{a}}(\mathbf{x})$, with $\epsilon = (\epsilon_1, \ldots, \epsilon_n)$ such that $0 \le \epsilon_i \le k+i-1$. Indeed, any nontrivial vanishing linear combination of the polynomials in $\mathcal{B}_{\mathbf{a}}$ would yield that a nontrivial linear combination of the monomials in $\mathcal{B}_{\mathbf{a}}$ vanishes modulo the ideal $I_{\mathbf{a}}$, thereby contradicting that these monomials are a basis for the quotient $\mathbb{Q}[\mathbf{x}]/I_{\mathbf{a}}$. This observation has the following consequence.

Theorem 8.5. *Writing* $\mathbf{a}$ *for* $\mathbf{k}^n + \delta_n$, *there is a natural isomorphism of graded* $\mathfrak{S}_n$-*modules between the space* $R = \mathbb{Q}[\mathbf{x}]$ *of polynomials and the tensor product* $R/I_{\mathbf{a}} \otimes \mathbb{Q}[h_{k+1}, \ldots, h_{k+n}]$. *In other words, every polynomial* $f(\mathbf{x})$ *in* R *has a unique expansion of the form* $\sum_{b \in \mathcal{B}_{\mathbf{a}}} f_b b(\mathbf{x})$, *where the* f_b *polynomials are in the* h_m *with* $k + 1 \le m \le k + n$.

Proof: Since $I_{\mathbf{a}}^\perp = \mathcal{H}_{\mathbf{a}}$ implies that any basis of $\mathcal{H}_{\mathbf{a}}$ can be used as a basis of the quotient $\mathbb{Q}[\mathbf{x}]/I_{\mathbf{a}}$, the above observation shows that we can find scalars c_b and polynomials $g_i(\mathbf{x})$ such that

$$f(\mathbf{x}) = \sum_{b \in \mathcal{B}_{\mathbf{a}}} c_b b(\mathbf{x}) + \sum_{i=1}^{n} g_i(\mathbf{x}) h_{k+i}(\mathbf{x}).$$

Iterating this decomposition for the g_i, we eventually conclude that the collection $\{b(\mathbf{x}) h_{k+1}^{d_1} \cdots h_{k+n}^{d_n}\}_{b \in \mathcal{B}_n; d_i \ge 0}$ spans the polynomial ring $\mathbb{Q}[\mathbf{x}]$. However, the generating function of their degrees is clearly given by the expression

$$F(t) = \prod_{i=1}^{n} \frac{1}{1 - t^{k+i}} \sum_{b \in \mathcal{B}_n} t^{\deg(b)}.$$

The straightforward degree enumeration of $\mathcal{B}_{\mathbf{a}}$ gives

$$\sum_{b \in \mathcal{B}_{\mathbf{a}}} t^{\deg(b)} = t^{kn + \binom{n}{2}} \prod_{i=1}^{n} \frac{1 - t^{-k-i}}{1 - t^{-1}} = \prod_{i=1}^{n} \frac{1 - t^{k+i}}{1 - t},$$

so that $F(t) = (1 - t)^{-n}$. Since the latter is precisely the Hilbert series of the polynomial ring $\mathbb{R}[\mathbf{x}]$, we are forced to conclude that the relevant collection is an independent set, thus completing the proof. $\square$

We are now in a position to prove the following special case of Theorem 8.1.

Theorem 8.6. *The graded Frobenius characteristic of* $\mathcal{H}_{\mathbf{k}^n + \delta_n}$ *is given by the polynomial*

$$\begin{bmatrix} k + n \\ n \end{bmatrix}_t \prod_{j=1}^{n} (1 - t^j) h_n \left[\frac{\mathbf{z}}{1 - t} \right]. \tag{8.8}$$

Proof: The uniqueness of the expansion given by Theorem 8.5 and the invariance of the coefficients under the action of $\mathfrak{S}_n$ yield that the Frobenius characteristics of $\mathcal{H}_{\mathbf{k}^n + \delta_n}$ and $R = \mathbb{Q}[\mathbf{x}]$ are related by the identity

$$\mathrm{Frob}_t(R) = \mathrm{Frob}_t(\mathcal{H}_{\mathbf{k}^n + \delta_n}) \prod_{j=1}^{n} \frac{1}{1 - t^{k+j}}.$$

Since we have seen that $\mathrm{Frob}_t(R) = h_n[\mathbf{z}/1 - t]$, from (8.4) we deduce that

$$\mathrm{Frob}_t(\mathcal{H}_{\mathbf{k}^n + \delta_n}) = \prod_{j=1}^{n} \frac{1 - t^{k+j}}{1 - t^j} \prod_{j=1}^{n} (1 - t^j) h_n\left[\frac{\mathbf{z}}{1 - t}\right],$$

whose right-hand side is just another way of writing (8.8). $\qquad\square$

Remark 8.7. We should mention that for $\mathbf{a} = \delta_n$, all these results are known. In particular, Theorem 8.5 is a generalization of Chevalley's Theorem (see (8.1)). In fact, most of the above statements generalize beautifully, and this is nicely discussed in the expository notes [Roth 05], where the following theorem is shown to hold for all (pseudo) reflection groups G.

Theorem 8.8 (Roth). *Suppose that V is a pseudo-reflection representation of G and $f_1, \ldots, f_m$ is a collection of polynomials spanning a G-invariant subspace of $\mathbb{R}[\mathbf{x}]$. Then $\mathcal{L}_\partial[f_1, \ldots, f_m]$ is a direct sum of copies of the regular representation of G if and only if it can be generated by G-skew invariant polynomials.*

Schur Differential Operators

As a byproduct of the above discussion, we get an interesting description of the effect of the operators $s_\mu(\partial\mathbf{x})$ on alternating polynomials of the form $\Delta_{\lambda+\delta_n}(\mathbf{x})$. This will be useful to have in mind in later considerations.

Proposition 8.9. *For a partition ν, let us set $d_\nu := \langle \Delta_{\nu+\delta_n}, \Delta_{\nu+\delta_n} \rangle$. Let λ and μ both be partitions with at most n parts. Then*

$$s_\mu(\partial\mathbf{x})\Delta_{\lambda+\delta_n}(\mathbf{x})/d_\lambda = \sum_{\nu \subseteq \lambda} c_{\mu,\nu}^\lambda \Delta_{\nu+\delta_n}(\mathbf{x})/d_\nu,$$

where $c_{\mu,\nu}^\lambda$ are the Littlewood–Richardson coefficients.

Proof: It follows from Lemma 8.2 that we have the expansion

$$s_\mu(\partial\mathbf{x})\Delta_{\lambda+\delta_n}(\mathbf{x}) = \sum_{\nu \vdash |\lambda| - |\mu|} \langle s_\mu(\partial\mathbf{x})\Delta_{\lambda+\delta_n}, \Delta_{\nu+\delta_n} \rangle \Delta_{\nu+\delta_n}(\mathbf{x})/d_\nu.$$

Since differentiation is dual to multiplication with respect to the scalar product, we see that we can write

$$\langle s_\mu(\partial \mathbf{x})\Delta_{\lambda+\delta_n}, \Delta_{\nu+\delta_n}\rangle = \langle \Delta_{\lambda+\delta_n}, s_\mu \Delta_{\nu+\delta_n}\rangle = \langle \Delta_{\lambda+\delta_n}, s_\mu s_\nu \Delta_n\rangle.$$

The Littlewood–Richardson rule then gives

$$\langle \Delta_{\lambda+\delta_n}, s_\mu s_\nu \Delta_n\rangle = \sum_{\theta \vdash |\lambda|} c_{\mu\nu}^\theta \langle \Delta_{\lambda+\delta_n}, s_\theta \Delta_n\rangle = \sum_{\theta \vdash |\lambda|} c_{\mu\nu}^\theta \langle \Delta_{\lambda+\delta_n}, \Delta_{\theta+\delta_n}\rangle,$$

and the orthogonality reduces this to

$$\langle \Delta_{\lambda+\delta_n}, s_\mu s_\nu \Delta_n\rangle = c_{\mu\nu}^\lambda \langle \Delta_{\lambda+\delta_n}, \Delta_{\lambda+\delta_n}\rangle.$$

Combining these identities yields

$$s_\mu(\partial \mathbf{x})\Delta_{\lambda+\delta_n}(\mathbf{x}) = \sum_{\nu \subseteq \lambda} c_{\mu,\nu}^\lambda \frac{d_\lambda}{d_\nu}\Delta_{\nu+\delta_n}(\mathbf{x}),$$

which gives the required equality upon division by d_λ.   □

8.7   Tensorial Square

For reasons that will be made clearer in Chapter 10, we now describe in detail the special case of (8.1) corresponding to the group $\mathfrak{S}_n \times \mathfrak{S}_n$ acting on the tensor square $\mathbb{R}[\mathbf{x}] \otimes \mathbb{R}[\mathbf{x}]$. We identify this last space with the ring $\mathcal{R} := \mathbb{R}[\mathbf{x}, \mathbf{y}]$ of polynomials in two n-sets of variables $\mathbf{x}$ and $\mathbf{y}$, with the first copy of $\mathfrak{S}_n$ acting as usual on the $\mathbf{x}$ variables and the second copy on the $\mathbf{y}$ variables. The originality here is to consider the isomorphism in (8.1) from the point of view of the restriction to the *diagonal subgroup* $\{(\sigma, \sigma) \mid \sigma \in \mathfrak{S}_n\}$ of $\mathfrak{S}_n \times \mathfrak{S}_n$, which is naturally identified with $\mathfrak{S}_n$. Observe that from this perspective, the ring of invariants of $\mathfrak{S}_n \times \mathfrak{S}_n$ identifies with $\mathbb{R}[\mathbf{x}]^{\mathfrak{S}_n} \otimes \mathbb{R}[\mathbf{y}]^{\mathfrak{S}_n}$. The associated coinvariant space is of dimension $n!^2$, and as a $\mathfrak{S}_n \times \mathfrak{S}_n$-module, it is simply the tensor product of the coinvariant space of $\mathfrak{S}_n$ with itself. However its structure becomes more interesting when it is considered as a $\mathfrak{S}_n$-module under the *diagonal action* of the symmetric group $\mathfrak{S}_n$. Unfolding the point of view outlined above, this is the linear and multiplicative action (said to be diagonal because of its similar effect on both of these sets of variables) for which $\sigma \cdot x_i = x_{\sigma(i)}$ and $\sigma \cdot y_i = y_{\sigma(i)}$. Likewise, we consider the space $\mathcal{H}_{\mathfrak{S}_n \times \mathfrak{S}_n}$ as an $\mathfrak{S}_n$-module under the diagonal action. This space naturally contains the now familiar space of $\mathfrak{S}_n$-diagonal harmonics $\mathcal{D}_n$, which will be the main subject of Chapter 10. A more detailed understanding of $\mathcal{H}_{\mathfrak{S}_n \times \mathfrak{S}_n}$ will shed light on the structure of $\mathcal{D}_n$. The point here is that $\mathcal{H}_{\mathfrak{S}_n \times \mathfrak{S}_n}$ (or equivalently $\mathcal{R}_{\mathfrak{S}_n \times \mathfrak{S}_n}$) is much easier to study than $\mathcal{D}_n$. In particular, it is easy to see that $\mathcal{H}_{\mathfrak{S}_n \times \mathfrak{S}_n} \simeq \mathcal{H}_n \otimes \mathcal{H}_n$, as $\mathfrak{S}_n$-modules, with the diagonal action of $\mathfrak{S}_n$ on the right-hand side.

Bigraded Actions and Bigraded Frobenius Characteristic

We naturally extend to this diagonal context most of the notions that have been considered for the space $\mathbb{R}[\mathbf{x}]$ (see Section 3.1). Observe that the diagonal action of $\mathfrak{S}_n$ on $\mathcal{R} = \mathbb{R}[\mathbf{x}, \mathbf{y}]$ preserves bidegree. Just as in the simply graded case, consider the projection $\pi_{mk} \colon \mathcal{R} \longrightarrow \mathcal{R}_{mk}$ on the *bihomogeneous* component $\mathcal{R}_{mk}$ of bidegree (m, k) of $\mathcal{R}$. In general, a subspace W of $\mathcal{R}$ is said to be bihomogeneous if, for all m and k, we have $W_{mk} := \pi_{m,k} W \subseteq W$. Moreover, when W is $\mathfrak{S}_n$-invariant, each W_{mk} in the bigraded decomposition $W = \bigoplus_{m,k \geq 0} W_{mk}$ must also be $\mathfrak{S}_n$-invariant. We can thus introduce the notion of the *bigraded Frobenius characteristic*,

$$\mathrm{Frob}_{qt}(W) := \sum_{m,k \geq 0} \mathcal{F}(W_{mk}) q^k t^m,$$

of such a space. Similarly, the *bigraded Hilbert series* of W is

$$\mathrm{Hilb}_{qt}(W) := \sum_{m \geq 0} \dim(W_{mk}) q^k t^m.$$

From our previous explicit description of $\mathcal{H}_n$, it follows that

$$\mathcal{H}_{\mathfrak{S}_n \times \mathfrak{S}_n} = \mathcal{L}_\partial [\Delta_n(\mathbf{x}) \Delta_n(\mathbf{y})].$$

Then the above discussion reveals that the (bigraded) Hilbert series of $\mathcal{H}_{\mathfrak{S}_n \times \mathfrak{S}_n}$ is

$$\mathrm{Hilb}_{qt}(\mathcal{H}_{\mathfrak{S}_n \times \mathfrak{S}_n}) = \mathrm{Hilb}_q(\mathcal{H}_{\mathfrak{S}_n}) \mathrm{Hilb}_t(\mathcal{H}_{\mathfrak{S}_n}), \tag{8.9}$$

which is a bigraded analog of $n!^2$. This is also clearly the case for the coinvariant space $\mathcal{R}_{\mathfrak{S}_n \times \mathfrak{S}_n}$. Using the fact that for the diagonal action of $\mathfrak{S}_n$ on $\mathcal{V} \otimes \mathcal{W}$ we have $\mathrm{Frob}(\mathcal{V} \otimes \mathcal{W}) = \mathrm{Frob}(\mathcal{V}) * \mathrm{Frob}(\mathcal{W})$, we easily obtain the following expression for the bigraded Frobenius characteristic, here denoted by $F_n(\mathbf{z}; q, t)$, of $\mathcal{H}_{\mathfrak{S}_n \times \mathfrak{S}_n}$,

$$F_n(\mathbf{z}; q, t) := \mathrm{Frob}_{qt}(\mathcal{H}_{\mathfrak{S}_n \times \mathfrak{S}_n})(\mathbf{z})$$
$$= \prod_{j=1}^n (1 - t^j)(1 - q^j) h_n \left[\frac{\mathbf{z}}{(1 - t)(1 - q)} \right]. \tag{8.10}$$

For example, we have

$$F_1(\mathbf{z}; q, t) = s_1,$$
$$F_2(\mathbf{z}; q, t) = (qt + 1)s_2 + (q + t)s_{11},$$
$$F_3(\mathbf{z}; q, t) = (q^3t^3 + q^2t^2 + q^2t + qt^2 + qt + 1)s_3,$$
$$+ (q^3t^2 + q^2t^3 + q^3t + q^2t^2 + qt^3$$
$$+ q^2t + qt^2 + q^2 + qt + t^2 + q + t)s_{21},$$
$$+ (q^2t^2 + q^3 + q^2t + qt^2 + t^3 + qt)s_{111}.$$

Writing the coefficients of these symmetric functions as matrices gives a better idea of the nice symmetries involved in these expressions. This is to say that the coefficient of $q^i t^j$ is the entry in position (i, j), starting[2] at $(0, 0)$ and going from bottom to top and left to right. Using this convention, $F_4(\mathbf{z}; q, t)$ equals

$$\begin{bmatrix} 0 & 0 & 0 & 0 & 0 & 0 & 1 \\ 0 & 0 & 0 & 1 & 1 & 1 & 0 \\ 0 & 0 & 1 & 1 & 2 & 1 & 0 \\ 0 & 1 & 1 & 2 & 1 & 1 & 0 \\ 0 & 1 & 2 & 1 & 1 & 0 & 0 \\ 0 & 1 & 1 & 1 & 0 & 0 & 0 \\ 1 & 0 & 0 & 0 & 0 & 0 & 0 \end{bmatrix} s_4 + \begin{bmatrix} 0 & 0 & 0 & 1 & 1 & 1 & 0 \\ 0 & 1 & 2 & 2 & 2 & 1 & 1 \\ 0 & 2 & 3 & 4 & 3 & 2 & 1 \\ 1 & 2 & 4 & 4 & 4 & 2 & 1 \\ 1 & 2 & 3 & 4 & 3 & 2 & 0 \\ 1 & 1 & 2 & 2 & 2 & 1 & 0 \\ 0 & 1 & 1 & 1 & 0 & 0 & 0 \end{bmatrix} s_{31} + \begin{bmatrix} 0 & 0 & 1 & 0 & 1 & 0 & 0 \\ 0 & 1 & 1 & 2 & 1 & 1 & 0 \\ 1 & 1 & 2 & 2 & 2 & 1 & 1 \\ 0 & 2 & 2 & 4 & 2 & 2 & 0 \\ 1 & 1 & 2 & 2 & 2 & 1 & 1 \\ 0 & 1 & 1 & 2 & 1 & 1 & 0 \\ 0 & 0 & 1 & 0 & 1 & 0 & 0 \end{bmatrix} s_{22}$$

$$+ \begin{bmatrix} 0 & 1 & 1 & 1 & 0 & 0 & 0 \\ 1 & 1 & 2 & 2 & 2 & 1 & 0 \\ 1 & 2 & 3 & 4 & 3 & 2 & 0 \\ 1 & 2 & 4 & 4 & 4 & 2 & 1 \\ 0 & 2 & 3 & 4 & 3 & 2 & 1 \\ 0 & 1 & 2 & 2 & 2 & 1 & 1 \\ 0 & 0 & 0 & 1 & 1 & 1 & 0 \end{bmatrix} s_{211} + \begin{bmatrix} 1 & 0 & 0 & 0 & 0 & 0 & 0 \\ 0 & 1 & 1 & 1 & 0 & 0 & 0 \\ 0 & 1 & 2 & 1 & 1 & 0 & 0 \\ 0 & 1 & 1 & 2 & 1 & 1 & 0 \\ 0 & 0 & 1 & 1 & 2 & 1 & 0 \\ 0 & 0 & 0 & 1 & 1 & 1 & 0 \\ 0 & 0 & 0 & 0 & 0 & 0 & 1 \end{bmatrix} s_{1111}.$$

We can easily reformulate equation (8.10) as

$$F_n(\mathbf{z}; q, t) = \sum_{\lambda \vdash n} f^\lambda(q, t) s_\lambda(\mathbf{z})$$
$$= \sum_{\lambda \vdash n} \prod_{j=1}^{n} (1 - t^j)(1 - q^j) s_\lambda \left[\frac{1}{(1 - q)(1 - t)} \right] s_\lambda(\mathbf{z}). \tag{8.11}$$

Here $f_\lambda(q, t)$ gives an explicit expression for the bigraded enumeration of λ-indexed *isotypic components* of $\mathcal{H}_{\mathfrak{S}_n \times \mathfrak{S}_n}$, i.e., all the irreducible compo-

---

[2]Using cartesian coordinates.

nents that are isomorphic to a given irreducible representation. In particular, the respective bigraded dimensions of the spaces $\mathcal{T}_n$, of *diagonally symmetric* harmonic polynomials, and $\mathcal{A}_n$, of *diagonally antisymmetric* harmonic polynomials, are

$$\mathrm{Hilb}_{qt}(\mathcal{T}_n) = f^{(n)}(q,t) = \sum_{\sigma \in \mathfrak{S}_n} q^{\mathrm{maj}(\sigma)} t^{\mathrm{maj}(\sigma^{-1})}, \qquad (8.12)$$

and

$$\mathrm{Hilb}_{qt}(\mathcal{A}_n) = f^{1^n}(q,t) = \sum_{\sigma \in \mathfrak{S}_n} q^{\mathrm{maj}(\sigma)} t^{\binom{n}{2}-\mathrm{maj}(\sigma^{-1})}. \qquad (8.13)$$

Here we have made use of results regarding $\mathrm{maj}(\sigma)$ that can be found in [Stanley 79]. Observe that we have the specializations

$$f^\lambda(q,1) = f^\lambda \prod_{k=1}^n \frac{q^k-1}{q-1}, \qquad f^\lambda(1,t) = f^\lambda \prod_{k=1}^n \frac{t^k-1}{t-1},$$
$$f^\lambda(q,0) = f^\lambda, \qquad f^\lambda(0,t) = f^\lambda.$$

There are many nice symmetries of the bigraded Frobenius characteristic $F_n(\mathbf{z};q,t)$. Noteworthy among these are

(1) $F_n(\mathbf{z};t,q) = F_n(\mathbf{z};q,t)$,

(2) $F_n(\mathbf{z};t,q) = (qt)^{\binom{n}{2}} F_n(\mathbf{z};q^{-1},t^{-1})$,

(3) $\omega\big(F_n(\mathbf{z};q,t)\big) = q^{\binom{n}{2}} F_n(\mathbf{z};q^{-1},t) = t^{\binom{n}{2}} F_n(\mathbf{z};q,t^{-1})$.

All of these symmetries correspond to automorphisms (or antiautomorphisms) of $\mathcal{H}_{\mathfrak{S}_n \times \mathfrak{S}_n}$. The first corresponds to the evident symmetry of $\mathcal{H}_{\mathfrak{S}_n \times \mathfrak{S}_n}$ that consists of exchanging the $\mathbf{x}$ variables with the $\mathbf{y}$ variables. The second corresponds to the morphism that sends a polynomial $f(\mathbf{x},\mathbf{y})$ to $f(\partial\mathbf{x},\partial\mathbf{y})\Delta_n(\mathbf{x})\Delta_n(\mathbf{y})$. The third corresponds to the *sign twisting* antiautomorphism that sends $f(\mathbf{x},\mathbf{y})$ to $f(\partial\mathbf{x},\mathbf{y})\Delta_n(\mathbf{x})$, for the first identity, or to $f(\mathbf{x},\partial\mathbf{y})\Delta_n(\mathbf{y})$, for the second. Further details and explicit descriptions of isotypic components of $\mathcal{H}_{\mathfrak{S}_n \times \mathfrak{S}_n}$ can be found in [Bergeron and Lamontagne 07]. An extension of this discussion to hyperoctahedral groups can also be found in [Bergeron and Biagioli 06].

Chapter 9

Macdonald Symmetric Functions

Looking for a common generalization of most of the important families of symmetric functions, Macdonald was led to introduce (in 1988, see [Macdonald 88]) a new family of two-parameter symmetric functions. His original definition made use of a common property of Schur, Zonal, Jack, and Hall–Littlewood symmetric functions, as well as new, more general symmetric functions that had just been considered by Kevin Kadell in his study of q-Selberg integrals. The common feature of all of these linear bases is that they can all be obtained by the same orthogonalization process, for a suitable choice of scalar product. Thus, we are naturally led to consider a more general scalar product that can be specialized to each of these individual cases. One further exciting feature of Macdonald's proof of the existence and uniqueness of his new symmetric functions was the introduction of self-adjoint operators for which they are common eigenfunctions with distinct eigenvalues. These operators would soon play a crucial role in the solution of the Calogero–Sutherland model of quantum many-body systems in statistical physics, instantly making Macdonald symmetric functions (polynomials) a fundamental part of this theory. It is not surprising that [Macdonald 95, Chapter VI] is the reference of choice here. A general review of recent work can also be found in [Garsia and Remmel 05].

9.1 Macdonald's Original Definition

Our context now is the ring $\mathbb{C}[\mathbf{x}]_{qt}^{\mathfrak{S}}$ of symmetric functions in an infinite[1] set of variables $\mathbf{z} = z_1, z_2, z_3, \ldots$, with coefficients in the field $\mathbb{C}(q, t)$ of rational fractions in two formal parameters, q and t. In terms of the power

---

[1]Or at least more than the degree of the polynomials considered.

sum basis $\{p_\lambda\}_\lambda$, we introduce on this ring the scalar product

$$\langle p_\lambda, p_\mu \rangle_{q,t} = \begin{cases} Z_\lambda(q,t) & \text{if } \lambda = \mu, \\ 0 & \text{otherwise,} \end{cases} \tag{9.1}$$

where

$$Z_\lambda(q,t) = z_\lambda \prod_{i=1}^{\ell(\lambda)} \frac{1-q^{\lambda_i}}{1-t^{\lambda_i}}. \tag{9.2}$$

In his original 1988 paper, Macdonald established the existence and uniqueness of symmetric functions (polynomials) $P_\mu = P_\mu(\mathbf{z}; q, t)$ satisfying the
following:

(1) $P_\mu = m_\mu + \sum_{\lambda \prec \mu} \xi_{\mu\lambda}(q,t) m_\lambda$, with coefficients $\xi_{\mu\lambda}(q,t)$ in $\mathbb{C}(q,t)$;

(2) $\langle P_\lambda, P_\mu \rangle_{q,t} = 0$, whenever $\lambda \neq \mu$.

Recall that $\lambda \prec \mu$ stands for λ being smaller than μ in the dominance
order. For example, we have

$$
\begin{aligned}
P_{31}(\mathbf{z}; q, t) = {} & \frac{(q+1)(t-1)^2(3q^2t+q^2+2qt+2q+t+3)}{24(qt-1)^2(qt+1)} p_1(\mathbf{z})^4 \\
& + \frac{(t-1)(t+1)(q^3t+q^3-q^2t+q^2-qt+q-t-1)}{4(qt-1)^2(qt+1)} p_2(\mathbf{z}) p_1(\mathbf{z})^2 \\
& + \frac{(q-1)(q+1)(t^2+t+1)(q-t)}{3(qt-1)^2(qt+1)} p_3(\mathbf{z}) p_1(\mathbf{z}) \\
& - \frac{(q-1)(q^2+1)(t-1)(t+1)^2}{8(qt-1)^2(qt+1)} p_2(\mathbf{z})^2 \\
& - \frac{(q+1)(q-1)^2(t+1)(t^2+1)}{4(qt-1)^2(qt+1)} p_4(\mathbf{z}).
\end{aligned}
$$

$$\tag{9.3}$$

Here $P_{31}(\mathbf{z})$ is expressed in terms of the power sum basis in order to make
apparent a natural simplification that will be the starting point of our
next section. Let us make a few observations. When $q = t$, the scalar
product (9.1) clearly specializes to the usual scalar product for symmetric
functions, hence we get $P_\mu(\mathbf{z}; q, q) = s_\mu(\mathbf{z})$. The Hall–Littlewood symmetric functions are obtained by setting $q = 0$, and the Jack symmetric
functions[2] by setting $q = t^\alpha$ ($\alpha \in \mathbb{R}$ and $\alpha > 0$) and letting $t \to 1$. Zonal

---

[2]For definitions and more properties of Hall–Littlewood, Jack and Zonal symmetric
functions, see [Macdonald 95]. There is no need here to go into more details; we only
mention these specializations to show that Macdonald functions play a unifying role.

symmetric functions are just the special case when $\alpha = 2$. Macdonald further observes that the symmetry $P_\mu(\mathbf{z}; q^{-1}, t^{-1}) = P_\mu(\mathbf{z}; q, t)$ follows from a similar symmetry of the scalar product (up to a constant factor). In particular, this implies that in the expansion

$$P_\mu(\mathbf{z}, q, t) = \sum_{\lambda \vdash n} \frac{\alpha_{\mu\lambda}(q, t)}{\beta_{\mu\lambda}(q, t)} p_\lambda(\mathbf{z})$$

of P_μ in terms of power sums, the coefficients (here expressed as reduced quotients of two polynomials in $\mathbb{Z}[q, t]$) must be such that $\deg_q(\alpha_{\mu\lambda}(q, t)) = \deg_q(\beta_{\mu\lambda}(q, t))$ and $\deg_t(\alpha_{\mu\lambda}(q, t)) = \deg_t(\beta_{\mu\lambda}(q, t))$. More surprisingly, we have the beautiful specializations:

$$P_\mu(\mathbf{z}; q, 1) = m_\mu \quad \text{and} \quad P_\mu(\mathbf{z}; 1, t) = e_{\mu'}.$$

9.2 Integral Renormalization

Throughout the rest of our discussion, we will consider a variation of Macdonald functions P_μ, denoted by H_μ and called the *integral form Macdonald functions*.[3] They are most easily described by a plethystic "change of variables" (followed by a renormalization)

$$H_\mu(\mathbf{z}; q, t) := P_\mu\left[\frac{\mathbf{z}}{1-t}; q, t^{-1}\right] \prod_{c \in \mu} (q^{a(c)} - t^{\ell(c)+1}). \qquad (9.4)$$

Here $a(c)$ and $\ell(c)$ are respectively the arm length and the leg length of the cell c of μ. The reason we can think of this as a simplification is that it transforms expressions like (9.3) into simpler expressions like the following:

$$H_{31}(\mathbf{z}; q, t) = \frac{1}{24}(q + 1)(q^2 t + 3q^2 + 2qt + 2q + 3t + 1)p_1^4$$
$$- \frac{1}{4}(q^3 t + q^3 + q^2 t - q^2 + qt - q - t - 1)p_2 p_1^2$$
$$+ \frac{1}{3}(qt - 1)(q + 1)(q - 1)p_3 p_1 + \frac{1}{8}(q^2 + 1)(q - 1)(t - 1)p_2^2$$
$$- \frac{1}{4}(q + 1)(q - 1)^2(t - 1)p_4.$$

In fact the elegance of the resulting symmetric functions is even more apparent when one expands them in terms of Schur functions.

---

[3]A word of caution about notation. Contrary to the convention in [Macdonald 95], also followed by others, we use H_μ instead of $\tilde{H}_\mu$ because we have no need of the alternate notion.

To illustrate, for $n = 2$ we have $H_2 = s_2 + qs_{11}$ and $H_{11} = s_2 + ts_1$; for $n = 3$,

$$H_3 = s_3 + (q^2 + q)s_{21} + q^3 s_{111},$$
$$H_{21} = s_3 + (q + t)s_{21} + qts_{111},$$
$$H_{111} = s_3 + (t^2 + t)s_{21} + t^3 s_{111};$$

and for $n = 4$,

$$H_4 = s_4 + (q^3 + q^2 + q)s_{31} + (q^4 + q^2)s_{22} + (q^5 + q^4 + q^3)s_{211}$$
$$+ q^6 s_{1111},$$
$$H_{31} = s_4 + (q^2 + q + t)s_{31} + (q^2 + qt)s_{22} + (q^3 + q^2 t + qt)s_{211}$$
$$+ q^3 ts_{1111},$$
$$H_{22} = s_4 + (qt + q + t)s_{31} + (q^2 + t^2)s_{22} + (q^2 t + qt^2 + qt)s_{211}$$
$$+ q^2 t^2 s_{1111},$$
$$H_{211} = s_4 + (q + t + t^2)s_{31} + (qt + t^2)s_{22} + (qt + qt^2 + t^3)s_{211} + qt^3 s_{1111},$$
$$H_{1111} = s_4 + (t + t^2 + t^3)s_{31} + (t^2 + t^4)s_{22} + (t^3 + t^4 + t^5)s_{211} + t^6 s_{1111}.$$

The coefficients $K_{\lambda,\mu}(q,t)$ that appear in the expansion

$$H_\mu(\mathbf{z}; q, t) = \sum_{\lambda \vdash n} K_{\lambda,\mu}(q,t)s_\lambda(\mathbf{z}) \tag{9.5}$$

of the H_μ in terms of the s_λ are called q,t-*Kostka polynomials*, in light of the results stated below. These coefficients are evidently equal to $\langle s_\lambda, H_\mu \rangle$. Observe that in each of our examples the value $K_{\lambda,\mu}(1,1)$ depends only on μ and n. We may also observe in these instances that $K_{\lambda,\mu}(0,1) = K_{\lambda,\mu}$ (the Kostka numbers, see (4.17)), so that in other terms we have

$$H_4|_{q=0,t=1} = s_4 = h_4,$$
$$H_{31}|_{q=0,t=1} = s_4 + s_{31} = h_{31},$$
$$H_{22}|_{q=0,t=1} = s_4 + s_{31} + s_{22} = h_{22},$$
$$H_{211}|_{q=0,t=1} = s_4 + 2s_{31} + s_{22} + s_{211} = h_{311},$$
$$H_{1111}|_{q=0,t=1} = s_4 + 3s_{31} + 2s_{22} + 3s_{211} + s_{1111} = h_{1111}.$$

This phenomenon (which will be confirmed in the next section) allows us to consider (9.5) as a q,t-analog of (4.17).

It was shown independently by several authors ([Garsia and Tesler 96], [Kirillov and Noumi 98], [Knop 97, Sahi 96], and [Lapointe and Vinet 97]), all in 1996, that actually

$$K_{\lambda,\mu}(q,t) \in \mathbb{Z}[q,t].$$

This gave a partial solution to Macdonald's 1988 conjecture, which stated that these are actually polynomials with positive integer coefficients, i.e., $K_{\lambda,\mu}(q,t) \in \mathbb{N}[q,t]$ (see [Macdonald 88]). Using the vocabulary introduced in Section 4.8, we may say that the Macdonald functions H_μ are Schur-positive. It took five more years to get a proof of the positivity part. The proof, due to Haiman, used the fact that the isospectral Hilbert scheme of points in the plane is normal, Cohen–Macaulay, and Gorenstein. The problem of finding an elementary proof is still open. For the moment, we will discuss the many symmetries that stand out in the expansion of the H_μ in terms of Schur functions. We will also give alternate descriptions of the H_μ.

9.3   Basic Formulas

Many identities follow directly (sometimes by passing to a limit) from a straightforward translation of results of Macdonald[4] in terms of this renormalized version of the P_μ. Rather than reproducing his proofs here, we will insist on the significance of the resulting properties of the functions $H_\mu(\mathbf{z}; q, t)$. The first three properties are the simplest,

$$\text{(a) } H_\mu(\mathbf{z}; 0, 0) = s_n, \quad \text{(b) } H_\mu(\mathbf{z}; 0, 1) = h_\mu, \quad \text{(c) } H_\mu(\mathbf{z}; 1, 1) = s_1^n, \quad (9.6)$$

with the last one holding a special place in our story. Reformulated in terms of q, t-Kostka polynomials, these equalities state that

$$\text{(a) } K_{\lambda,\mu}(0,0) = \delta_{\lambda,n}, \quad \text{(b) } K_{\lambda,\mu}(0,1) = K_{\lambda,\mu}, \quad \text{(c) } K_{\lambda,\mu}(1,1) = f^\lambda. \tag{9.7}$$

Passing over the first, we see that the second of these justifies our terminology for q, t-Kostka polynomials. The last one has the striking property that it is independent of μ.

Exercise. Using the specializations of the P_μ mentioned in Section 9.1, show that the identities in (9.6) hold.

Observe that the right-hand side of (9.6)(c) is the Frobenius characteristic of the regular representation. This is one of the observations that led Garsia and Haiman to propose a representation-theoretic interpretation for $H_\mu(\mathbf{z}; q, t)$, which we will discuss in Section 10.1. For the time being, let us go on with our observations. The next properties that stand out are the symmetries

$$H_{\mu'}(\mathbf{z}; q, t) = H_\mu(\mathbf{z}; t, q), \tag{9.8}$$

$$H_\mu(\mathbf{z}; q, t) = q^{n(\mu')} t^{n(\mu)} \omega\big(H_\mu(\mathbf{z}; q^{-1}, t^{-1})\big), \tag{9.9}$$

---

[4] See [Macdonald 88] for more details.

which are also easily derived from the symmetries of the P_μ. Once reformulated in terms of q, t-Kostka polynomials, identities (9.8) and (9.9) take the form

$$K_{\lambda,\mu}(q,t) = K_{\lambda,\mu'}(t,q), \tag{9.10}$$

$$K_{\lambda,\mu}(q,t) = q^{n(\mu')}t^{n(\mu)}K_{\lambda',\mu}(q^{-1},t^{-1}). \tag{9.11}$$

These can easily be stated in terms of the q, t-Kostka matrix, which for $n = 4$ is

$$\begin{pmatrix} 1 & q^3 + q^2 + q & q^4 + q^2 & q^5 + q^4 + q^3 & q^6 \\ 1 & q^2 + q + t & q^2 + qt & q^3 + q^2t + qt & q^3t \\ 1 & qt + q + t & q^2 + t^2 & q^2t + qt^2 + qt & t^2q^2 \\ 1 & q + t^2 + t & qt + t^2 & qt^2 + qt + t^3 & qt^3 \\ 1 & t^3 + t^2 + t & t^4 + t^2 & t^5 + t^4 + t^3 & t^6 \end{pmatrix}.$$

Each row of the matrix corresponds to one of the H_μ, ordered in decreasing lexicographic order: 4, 31, 22, 211, 1111. The columns (likewise ordered) correspond to coefficients of Schur functions in these H_μ. The first column thus corresponds to the coefficient of s_n, so (9.7)(a) explains the 1s. All entries of the λ-indexed column share the same specialization at $q = t = 1$, which is the number f^λ of standard tableaux of shape λ. Reflecting with respect to the middle row and exchanging q and t gives a symmetry that corresponds to (9.10). A reflection with respect to the middle column is behind the similar but more intricate symmetry of (9.11). This phenomenon holds for all n.

In light of these observations, it is tempting to look for a rule, parametrized by μ, that would associate to each standard tableau τ of shape λ some monomial $q^{\alpha(\tau,\mu)}t^{\beta(\tau,\mu)}$ in such a way that

$$K_{\lambda,\mu}(q,t) = \sum_\tau q^{\alpha(\tau,\mu)}t^{\beta(\tau,\mu)}. \tag{9.12}$$

We would like to find some integer statistics $\alpha(\tau,\mu)$ and $\beta(\tau,\mu)$ that would make equation (9.12) true. Except for some very special cases, this objective has not been reached. The only known combinatorial explanation for the positivity of the $K_{\lambda,\mu}(q,t)$ relies on the fact that they enumerate the bigraded multiplicities of some irreducible representations (see Chapter 10). This approach, due to Garsia and Haiman, has sparked many new considerations at the frontier of algebraic combinatorics, representation theory, and algebraic geometry.

We now give various other interesting specializations of the H_μ, all following directly from results of Macdonald (see [Macdonald 88]). As we will see later, each opens up interesting possibilities and is related to some combinatorial point of view. The first consists of setting $t = 1$ (similarly, the $q = 1$ case follows from previously mentioned symmetries). With this specialization, the functions H_μ become multiplicative,[5] meaning that

$$H_\mu(\mathbf{z}; q, 1) = H_{\mu_1}(\mathbf{z}; q, 1) \cdots H_{\mu_k}(\mathbf{z}; q, 1), \tag{9.13}$$

so that they are entirely characterized by the simple functions $H_n(\mathbf{z}; q, 1)$. These are particularly interesting since they coincide with the Frobenius characteristic of the $\mathfrak{S}_n$-coinvariant space. This is to say that we have

$$H_n(\mathbf{z}; q, 1) = h_n \left[\frac{\mathbf{z}}{1 - q} \right] \prod_{j=1}^{n} (1 - q^j). \tag{9.14}$$

A second interesting specialization comes from setting $t = q^{-1}$. We then get the following expansion in terms of Schur functions:

$$H_\mu(\mathbf{z}; q, q^{-1}) = q^{-n(\mu)} s_\mu \left[\frac{\mathbf{z}}{1 - q} \right] \prod_{c \in \mu} (1 - q^{a(c) + \ell(c) + 1}). \tag{9.15}$$

Evidently, the two specializations just considered agree in the case $\mu = (n)$, since t does not appear in the expansion of $H_{(n)}(\mathbf{z}; q, t)$. Other formulas are obtained by specializing $\mathbf{z}$ (using plethystic substitutions). For instance, we have

$$H_\mu[1 - u; q, t] = \prod_{(i,j) \in \mu} (1 - q^i t^j u). \tag{9.16}$$

This allows us to check that $K_{\lambda,\mu}(q, t) \in \mathbb{N}[q, t]$ for special values of λ. Indeed, expanding each side of (9.16), we get

$$\sum_\lambda K_{\lambda,\mu}(q, t) s_\lambda[1 - u] = (1 - u) \sum_{k=0}^{n-1} e_k[B_\mu - 1](-u)^k,$$

where, as in Section 1.3, $B_\mu = B_\mu(q, t) := \sum_{(i,j) \in \mu} q^i t^j$ (see Figure 9.1). Equation (4.26) gives

$$s_\lambda[1 - u] = \begin{cases} (-u)^k (1 - u) & \text{if } \lambda = (n - k, 1^k), \\ 0 & \text{otherwise,} \end{cases}$$

so we deduce that

$$K_{\lambda,\mu}(q, t) = e_k[B_\mu - 1] \tag{9.17}$$

whenever $\lambda = (n - k, 1^k)$ is a hook shape. It may be helpful here to point out that $e_k[B_\mu - 1]$ stands for the elementary symmetric function

---

[5]This follows from the fact that with the same substitution the P_μ become multiplicative, as shown by Macdonald.

$$t^3$$

$$t^2 \quad qt^2$$

$$t \quad qt \quad q^2t \quad q^3t$$

$$1 \quad q \quad q^2 \quad q^3$$

Figure 9.1. $B_{4421} - 1 = q^3t + q^3 + q^2t + qt^2 + t^3 + q^2 + qt + t^2 + q + t.$

$e_k(q, t, q^2, qt, t^2, \dots)$ in the "variables" $q^i t^j$, for (i, j) varying in the set of cells of μ excepting $(0, 0)$. Clearly the resulting polynomial has positive integer coefficients. Observe that in the case $k = 0$ (corresponding to $\lambda = (n)$) equation (9.17) gives back (9.7)(a). At the other extreme, it is easy to check that $K_{\mu 1^n} = q^{n(\mu')} t^{n(\mu)}$, which can also be deduced from (9.9). As a more typical example, consider the partition of Figure 9.1, where each cell (i, j) has been labeled $q^i t^j$. We get

$$
\begin{aligned}
K_{4421,911}(q, t) &= e_2(q^3t, q^3, q^2t, qt^2, t^3, q^2, qt, t^2, q, t) \\
&= q^6t + q^5t^2 + q^4t^3 + q^3t^4 \\
&\quad + 2q^5t + 2q^4t^2 + 3q^3t^3 + q^2t^4 + qt^5 \\
&\quad + q^5 + 3q^4t + 4q^3t^2 + 3q^2t^3 + 2qt^4 + t^5 \\
&\quad + q^4 + 3q^3t + 3q^2t^2 + 3qt^3 + t^4 \\
&\quad + q^3 + 2q^2t + 2qt^2 + t^3 \\
&\quad + qt.
\end{aligned}
$$

To close this section, let us reformulate the orthogonality property of the P_μ in the guise of a Cauchy formula for the H_μ:

$$
e_n\left[\frac{\mathbf{zw}}{(1-t)(1-q)}\right] = \sum_{\mu \vdash n} \frac{H_\mu(\mathbf{z})H_\mu(\mathbf{w})}{\varepsilon_\mu(q, t)\varepsilon'_\mu(q, t)}, \tag{9.18}
$$

where $\varepsilon_\mu(q, t)$ and $\varepsilon'_\mu(q, t)$ are the following analogs of hook length products:

$$
\varepsilon_\mu(q, t) := \prod_{c \in \mu} q^{a(c)} - t^{\ell(c)+1} \quad \text{and} \quad \varepsilon'_\mu(q, t) := \prod_{c \in \mu} t^{\ell(c)} - q^{a(c)+1}.
$$

9.4 Equations for q, t-Kostka Polynomials

Rather than continue with the approach sketched in the previous sections, we instead characterize the q, t-Kostka polynomials as the unique solution of the system of linear equations:

$$\langle s_\lambda, H_\mu[\mathbf{z}(1-q); q, t]\rangle = 0, \quad \lambda \not\trianglelefteq \mu, \tag{9.19}$$
$$\langle s_\lambda, H_\mu[\mathbf{z}(1-t); q, t]\rangle = 0, \quad \lambda \not\trianglelefteq \mu', \tag{9.20}$$
$$\langle s_n, H_\mu(\mathbf{z}; q, t)\rangle = 1. \tag{9.21}$$

For example, with $\mu = 21$ we get the equations

$$q^2(1-q)K_{3,21}(q,t) + q(1-q)K_{21,21}(q,t) + (1-q)K_{111,21}(q,t) = 0,$$
$$t^2(1-t)K_{3,21}(q,t) + t(1-t)K_{21,21}(q,t) + (1-t)K_{111,21}(q,t) = 0,$$
$$K_{3,21}(q,t) = 1.$$

Solving for the $K_{\lambda,\mu}$, we find that $K_{3,21}(q,t) = 1$, $K_{21,21}(q,t) = q + t$, and $K_{111,21}(q,t) = qt$, which coincide with our previous assertion that

$$H_{21} = s_3 + (q+t)s_{21} + qts_{111}.$$

Observe that these are also special instances of (9.17), so that we learn nothing new, but this is just because we chose an example simple enough to fit the page.

9.5   A Combinatorial Approach

Very recently, a combinatorial formula for the expansion of the H_μ in the basis of monomial symmetric functions has been proposed by J. Haglund. He shows in [Haglund et al. 05a] that this formula indeed satisfies the required equations (of Section 9.4). Thus, the H_μ can now be very simply described as a sum over all fillings of μ-shape tableaux of certain terms in the parameters q and t and variables $\mathbf{z}$. These terms are obtained by multiplying the evaluation monomial $\mathbf{z}_\tau$ (see Section 4.1) of a tableaux $\tau\colon \mu \to \mathbb{N}$ of shape μ by suitably chosen powers of q and t defined in terms of *tableau descents*,

$$\mathrm{Des}(\tau) := \{(i,j) \in \mu \mid \tau(i, j-1) < \tau(i,j)\}, \tag{9.22}$$

and "tableau inversions", defined as follows. A cell a is said to *attack* another cell b if either both are in the same row with b to the right of a; or if b is in the row below that of a with b to the left of a, i.e., $a = (i+1, j)$ and $b = (i, k)$, with $j > k$. In visual terms, the shaded cells in Figure 9.2 are those that are attacked by the dotted cell. The set $\mathrm{Inv}(\tau)$ of *tableau inversions* of a μ-shape tableau τ is the set of pairs of cells (a, b) of μ, with b attacked by a and $\tau(a) > \tau(b)$. The *inversion number* of the tableau τ is $\ell(\tau) := |\mathrm{Inv}(\tau)| - \sum_{c \in \mathrm{Des}(\tau)} a(c)$, and its *major index* is the sum

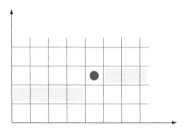

Figure 9.2. Attack region for the dotted cell.

$\mathrm{maj}(\tau) := \sum_{c \in \mathrm{Des}(\tau)} \ell(c) + 1$. We can now state Haglund's formula as follows:

$$H_\mu(\mathbf{z}; q, t) = \sum_{\tau \,:\, \mu \to \mathbb{N}} t^{\mathrm{maj}(\tau)} q^{\ell(\tau)} \mathbf{z}_\tau. \tag{9.23}$$

Let us illustrate with $\mu = 21$. For integers $a < b < c$, each tableau of shape 21 has one of the forms shown below.

$$
\begin{array}{cccccc}
z_a z_b z_c & q z_a z_b z_c & t z_a z_b z_c & t z_a z_b z_c & q z_a z_b z_c & q t z_a z_b z_c \\
\end{array}
$$

$$
\begin{array}{cccccc}
\boxed{a} & \boxed{a} & \boxed{c} & \boxed{b} & \boxed{b} & \boxed{c} \\
\boxed{b}\ \boxed{c} & \boxed{c}\ \boxed{b} & \boxed{a}\ \boxed{b} & \boxed{a}\ \boxed{c} & \boxed{c}\ \boxed{a} & \boxed{b}\ \boxed{a} \\
\end{array}
$$

$$
\begin{array}{ccccc}
z_a^2 z_b & q z_a^2 z_b & t z_a^2 z_b & t z_a z_b^2 & q z_a z_b^2 \\
\end{array}
$$

$$
\begin{array}{ccccc}
\boxed{a} & \boxed{a} & \boxed{b} & \boxed{b} & \boxed{b} \\
\boxed{a}\ \boxed{b} & \boxed{b}\ \boxed{a} & \boxed{a}\ \boxed{a} & \boxed{a}\ \boxed{b} & \boxed{b}\ \boxed{a} \\
\end{array}
$$

$$
z_a z_b^2 \tag{9.24}
$$

$$
\begin{array}{c}
\boxed{a} \\
\boxed{b}\ \boxed{b} \\
\end{array}
$$

$$
z_a^3
$$

$$
\begin{array}{c}
\boxed{a} \\
\boxed{a}\ \boxed{a} \\
\end{array}
$$

Summing over all possible choices of a, b, and c, we get the quasisymmetric monomial basis expression

$$H_{21}(\mathbf{z}; q, t) = M_3 + (1 + q + t) M_{21} + (1 + q + t) M_{12} + (1 + 2q + 2t + qt) M_{111},$$

which we can rewrite in terms of monomial symmetric functions as $m_3 + (1 + q + t) m_{21} + (1 + 2q + 2t + qt) m_{111}$, or in terms of Schur functions

$$
\begin{array}{cccc}
4 & 1 & & \\
3 & 6 & 2 & \\
5 & 7 & 8 & 9
\end{array}
$$

Figure 9.3. Permutational tableau.

as $s_3 + (q + t)s_{21} + qts_{111}$. When expressed in terms of the fundamental basis of quasisymmetric functions, equation (9.23) takes a particularly nice form.[6] For an n-cell diagram $\mathbf{d}$, a tableau $\pi \colon \mathbf{d} \to \{1, \ldots, n\}$ is said to be *permutational* if π is one-to-one. Thus, each number i between 1 and n appears once and only once in π, as illustrated in Figure 9.3. Any $\mathbf{d}$-shape tableau $\tau \colon \mathbf{d} \to \{1, 2, \ldots\}$ can be turned into a special permutational tableau, denoted by $\pi(\tau)$, by the following process. Suppose that τ contains m_i copies of the integer i. In reading order, we successively replace the 1s appearing in τ by the numbers $1, 2, \ldots, m_1$; then we replace in the same manner the 2s by the numbers $m_1 + 1, m_1 + 2, \ldots, m_1 + m_2$, etc. For example, for the tableau

$$
\begin{array}{cccc}
3 & 1 & & \\
2 & 4 & 1 & \\
3 & 4 & 4 & 4
\end{array}
$$

we get the permutational tableau of Figure 9.3, and in (9.24) the diagrams that correspond to the same permutational tableau appear in the same column. The point of this construction is that we have

$$
\mathrm{maj}(\tau) = \mathrm{maj}\big(\pi(\tau)\big) \quad \text{and} \quad \ell(\tau) = \ell\big(\pi(\tau)\big). \tag{9.25}
$$

In equation (9.26) below, we apply this π-statistic to special tableaux associated with pairs (σ, μ), with μ a given partition and σ any permutation in $\mathfrak{S}_n$. More precisely, we denote by $\pi(\sigma, \mu)$ the shape-μ permutational tableau obtained by simply filling the cells of μ with the values $\sigma_1, \sigma_2, \ldots, \sigma_n$, row by row, from top to bottom, and going from left to right within each row. Thus, for $\sigma = 413625789$ and $\mu = 432$, we get the tableau of Figure 9.3. Observe that the definition of descent set for permutations translates into

$$
\mathrm{Des}(\sigma^{-1}) = \{i \mid i + 1 \text{ is to the left of } i \text{ in } \sigma\},
$$

so that in view of (9.25), we can collect terms in Haglund's formula to get the following expansion in terms of the $Q_{\mathbf{a}}$:

$$
H_\mu(\mathbf{z}; q, t) = \sum_{\sigma \in \mathfrak{S}_n} t^{\mathrm{maj}(\pi(\sigma, \mu))} q^{\ell(\pi(\sigma, \mu))} Q_{\mathrm{co}(\mathrm{Des}(\sigma^{-1}))}. \tag{9.26}
$$

---

[6]We follow here a presentation of A. Garsia, made in a private communication to the author.

Indeed, the statement in (9.26) follows from the fact that

$$\sum_{\pi(\tau)=\pi(\sigma,\mu)} \mathbf{z}_\tau = Q_{\text{co}(\text{Des}(\sigma^{-1}))}.$$

Note that it is not obvious from (9.23) or (9.26) that H_μ is actually symmetric. This is proved in [Haglund et al. 05a] using a very nice expansion of the H_μ in terms of a family of Schur-positive symmetric functions originally considered in [Lascoux et al. 97].

9.6 The ∇ (Nabla) Operator

Recall that the integral form Macdonald functions H_μ are a linear basis of the ring of symmetric functions (with coefficients in the fraction field $\mathbb{C}(q,t)$). Thus, in order to describe a linear operator on this ring it is sufficient to state its effect on the H_μ. We can therefore define an operator ∇ by $\nabla(H_\mu) := q^{n(\mu')} t^{n(\mu)} H_\mu$, so that the H_μ are eigenfunctions of ∇ with simple, explicit eigenvalues. As we will shortly explain, the effect of ∇ on the elementary symmetric functions, here expressed in the basis of Schur functions, is striking:

$$\nabla(e_1) = s_1,$$

$$\nabla(e_2) = s_2 + (q+t)s_{11},$$

$$\nabla(e_3) = s_3 + (q^2 + qt + t^2 + q + t)s_{21} + (t^3 + qt^2 + q^2t + q^3 + qt)s_{111}.$$

In fact, the matrix of ∇ in the Schur function basis for some fixed degree n exhibits interesting unexpected features. To illustrate, let us set $\nabla_{\lambda\mu} := \langle \nabla(s_\lambda), s_\mu \rangle$. Then ∇ is defined for functions of degree n by the matrix $\nabla_n = (\nabla_{\lambda\mu})_{\lambda,\mu \vdash n}$, where we order the partitions in decreasing lexicographic order, e.g, 3, 21, 111, for $n = 3$. We have

$$\nabla_2 = \begin{bmatrix} 0 & -tq \\ 1 & t+q \end{bmatrix}$$

and

$$\nabla_3 = \begin{bmatrix} 0 & t^2q^2 & (t+q)t^2q^2 \\ 0 & -(t+q)tq & -(t^2+tq+q^2)tq \\ 1 & t^2+tq+q^2+t+q & t^3+t^2q+tq^2+q^3+tq \end{bmatrix}.$$

Careful inspection of such matrices for $2 \leq n \leq 6$, along with some theoretical considerations, led to the following conjecture [Bergeron et al. 99a]:

Conjecture 9.1 *For all λ and μ, the polynomial $(-1)^{\kappa(\lambda)}\langle \nabla(s_\lambda), s_\mu \rangle$ has positive integer coefficients, with $\kappa(\lambda)$ equal to*

$$\binom{k}{2} + \sum_{\lambda_i' < (i-1)} (i - 1 - \lambda_i').$$

Similar conjectures have been formulated in [Bergeron et al. 99a] for ∇^k and for more general operators. Conjecture 9.1 has been shown to hold in the special case $t = 1$ in [Lenart 00]. Among other work along these lines, we mention [Can and Loehr 06].

On another note, observe that there is a very simple expression of the inverse of ∇ in terms of ∇ itself. Let us denote by ω^* the operator[7] such that

$$\omega^* f(\mathbf{x}; q, t) = (\omega f[\mathbf{x}; q, t])_{\substack{q \mapsto 1/q \\ t \mapsto 1/t}}.$$

Then we easily check that the operator ∇ affords as inverse the expression $\nabla^{-1} = \omega^* \nabla \omega^*$. Indeed, equation (9.9) can be written as

$$\omega^* H_\mu[\mathbf{x}; q, t] = q^{-n(\mu')} t^{-n(\mu)} H_\mu[\mathbf{x}; q, t],$$

and thus

$$\begin{aligned}
\omega^* \nabla(\omega^* H_\mu[\mathbf{x}; q, t]) &= \omega^* \nabla(q^{-n(\mu')} t^{-n(\mu)} H_\mu[\mathbf{x}; q, t]) \\
&= \omega^* H_\mu[\mathbf{x}; q, t] \\
&= q^{-n(\mu')} t^{-n(\mu)} H_\mu[\mathbf{x}; q, t] \\
&= \nabla^{-1} H_\mu[\mathbf{x}; q, t].
\end{aligned}$$

It also follows that $\nabla \omega^*$ is an involution. The operator ∇ plays an important role in various contexts. Moreover, it is but one of a family of very interesting operators that have integral form Macdonald functions as common eigenvalues. Among these we find the operators ∇_f defined by

$$\nabla_f H_\mu := f[B_\mu] H_\mu,$$

where f is any symmetric function and $B_\mu := \sum_{(i,j) \in \mu} q^i t^j$. For more on all this see [Bergeron et al. 99a].

---

[7]Some care must be used in applying this operator. It is linear over the field of scalars $\mathbb{C}$, but not over $\mathbb{C}(q, t)$.

Chapter 10

Diagonal Coinvariant Spaces

In this chapter, we depart from our previous considerations in that we are now going to systematically consider the diagonal action (see Section 8.7) of the symmetric group $\mathfrak{S}_n$ on the ring $\mathcal{R} = \mathbb{C}[\mathbf{x}, \mathbf{y}]$ of polynomials in two sets of variables $\mathbf{x}$ and $\mathbf{y}$. This is not an action that corresponds to a reflection group action of $\mathfrak{S}_n$, so none of the results considered earlier can be directly applied. It develops that this new context is entirely different, both in the kinds of results that can be obtained and in the difficulty of proving them. It has been the object of intensive study in the last 15 years, and many new avenues of research have opened up as a result. For a more in-depth discussion see M. Haiman's very nice survey in [Haiman 01].

10.1 Garsia–Haiman Representations

While searching for an algebraic combinatorial proof of Macdonald's conjecture (Section 9.2), Garsia and Haiman (see [Garsia and Haiman 96b]) were led to consider the following generalization of the $\mathfrak{S}_n$-module of harmonic polynomials. For any partition μ of n, define the determinant

$$\Delta_\mu(\mathbf{x}, \mathbf{y}) := \det(x_k^j y_k^i)_{\substack{1 \le k \le n \\ (i,j) \in \mu}}. \tag{10.1}$$

In principle, this is defined only up to sign, since no explicit order for listing the cells of μ is mentioned here. For our general purpose, any order would be fine, but let us choose the lexicographic order. For instance, choosing $\mu = 211$, we get

$$\Delta_{211} = \det \begin{pmatrix} 1 & x_1 & y_1 & y_1^2 \\ 1 & x_2 & y_2 & y_2^2 \\ 1 & x_3 & y_3 & y_3^2 \\ 1 & x_4 & y_4 & y_4^2 \end{pmatrix} = x_1 y_3 y_2^2 - x_1 y_4 y_2^2 - x_1 y_2 y_3^2 + x_1 y_4 y_3^2 +$$

$$+ x_1y_2y_4^2 - x_1y_3y_4^2 - x_2y_3y_1^2 + x_2y_4y_1^2 + x_2y_1y_3^2$$
$$- x_2y_4y_3^2 - x_2y_1y_4^2 + x_2y_3y_4^2 + x_3y_2y_1^2 - x_3y_4y_1^2$$
$$- x_3y_1y_2^2 + x_3y_4y_2^2 + x_3y_1y_4^2 - x_3y_2y_4^2 - x_4y_2y_1^2$$
$$+ x_4y_3y_1^2 + x_4y_1y_2^2 - x_4y_3y_2^2 - x_4y_1y_3^2 + x_4y_2y_3^2.$$

Evidently, when $\mu = (n)$ we obtain the usual Vandermonde determinant in the variables $\mathbf{x}$, and when $\mu = 1^n$, the Vandermonde determinant in the variables $\mathbf{y}$. Together with Garsia and Haiman, we now consider the μ-*harmonic subspace* $\mathcal{H}_\mu := \{f(\partial\mathbf{x}, \partial\mathbf{y})\Delta_\mu(\mathbf{x}, \mathbf{y}) \mid f(\mathbf{x}, \mathbf{y}) \in \mathcal{R}\}$, which is the smallest subspace of $\mathcal{R}$ containing $\Delta_\mu(\mathbf{x}, \mathbf{y})$ and closed under partial derivatives. Clearly $\mathcal{H}_\mu$ is a bihomogeneous invariant subspace of $\mathcal{R}$. In the early 1990s, Garsia and Haiman formulated the conjecture that

$$\mathrm{Frob}_{qt}(\mathcal{H}_\mu) = H_\mu, \tag{10.2}$$

which provides an explicit description of the integral form Macdonald functions. This conjecture has a number of consequences. In particular, it implies that $\mathcal{H}_\mu$ is a bigraded version of the regular representation. Thus, it further implies that $\dim \mathcal{H}_\mu = n!$. In fact, Haiman showed (see [Garsia and Haiman 93]) early in the work on this problem that this seemingly simpler fact implies equation (10.2). This is where the research stood for many years, during which time the claim $\dim \mathcal{H}_\mu = n!$ became widely known as the $n!$-Conjecture. It was finally proved by Haiman in 2002, and so it is now known as the $n!$-Theorem.

Another consequence of statement (10.2) is that the q, t-Kostka polynomial $K_{\lambda,\mu}(q, t)$ can be interpreted as the bigraded multiplicity of the irreducible representation indexed by λ in the $\mathfrak{S}_n$-module $\mathcal{H}_\mu$. This in turn proves Macdonald's conjecture (see Section 9.2).

To illustrate, consider the partition $\mu = 21$, for which

$$\Delta_{21} = x_1y_2 - x_1y_3 - x_2y_1 + x_2y_3 + x_3y_1 - x_3y_2.$$

We can calculate the following bigraded basis of the space $\mathcal{H}_{21}$:

Basis element	bidegree	Frobenius characteristic
Δ_{21}	$(1, 1)$	s_{111}
$x_1 - x_2, x_1 - x_3$	$(1, 0)$	s_{21}
$y_1 - y_2, y_1 - y_3$	$(0, 1)$	s_{21}
1	$(0, 0)$	s_3.

Then the bigraded Frobenius characteristic of $\mathcal{H}_{21}$ is $\mathrm{Frob}_{qt}(\mathcal{H}_{21}) = s_3 + (q + t)s_{21} + qts_{111}$, which does coincide with the integral form Macdonald function H_{21}.

Many striking symmetries of integral form Macdonald functions can be understood directly in terms of the spaces $\mathcal{H}_\mu$. For instance, it is clear that $\Delta_\mu(\mathbf{x}, \mathbf{y}) = \Delta_{\mu'}(\mathbf{y}, \mathbf{x})$, hence a simple exchange of the variables $\mathbf{x}$ and $\mathbf{y}$ establishes an isomorphism of the spaces $\mathcal{H}_\mu$ and $\mathcal{H}_{\mu'}$. This translates into the equality $\mathrm{Frob}_{qt}(\mathcal{H}_\mu) = \mathrm{Frob}_{tq}(\mathcal{H}_{\mu'})$, which is exactly equation (9.8). Now, consider the μ-*flip* $\downarrow_\mu\colon \mathcal{H}_\mu \longrightarrow \mathcal{H}_\mu$, defined by $\downarrow_\mu p(\mathbf{x}, \mathbf{y}) := p(\partial\mathbf{x}, \partial\mathbf{y})\Delta_\mu(\mathbf{x}, \mathbf{y})$. This bijective linear transformation is *degree complementing*, i.e., $\beta(\downarrow_\mu p) = \beta(\Delta_\mu) - \beta(p)$, and *sign twisting*, i.e., $\sigma \cdot \downarrow_\mu p = \mathrm{sgn}(\sigma) \downarrow_\mu \sigma \cdot p$. Thus, we have the resulting identity $\omega\,\mathrm{Frob}_{qt}(\mathcal{H}_\mu) = q^{n(\mu')}t^{n(\mu)}\,\mathrm{Frob}_{q^{-1}t^{-1}}(\mathcal{H}_\mu)$, which agrees with (9.9). For example, applying this to the space $\mathcal{H}_{21}$ gives the following result:

Frob. char.	$p(\mathbf{x}, \mathbf{y})$	$\downarrow_\mu p(\mathbf{x}, \mathbf{y})$	Frob. char.
s_{111}	Δ_{21}	6	s_3
s_{21}	$x_1 - x_2, x_1 - x_3$	$y_1 + y_2 - 2y_3, -y_1 + 2y_2 - y_3$	s_{21}
s_{21}	$y_1 - y_2, y_1 - y_3$	$-x_1 - x_2 + 2x_3, x_1 - 2x_2 + x_3$	s_{21}
s_3	1	Δ_{21}	s_{111}

Partial Flag Cohomology Ring

Although this takes us into a realm that is a bit far from our current theme,[1] it is interesting to observe that the space $\mathcal{H}_\mu$ contains a space closely related to the cohomology ring of the μ-flag variety considered in [De Concini and Procesi 81]. De Concini and Procesi have given an elementary description of this cohomology ring, which makes it fall back into the kinds of objects that we have been dealing with. More precisely, they show that the cohomology ring in question is isomorphic to the space $\mathcal{G}_\mu$ spanned by all partial derivatives of the polynomials $\Delta_\tau(\mathbf{x})$ (recall Definition 5.5), for τ varying in the set of injective tableaux of shape μ. The precise link is that this space is exactly the set $\mathcal{H}_\mu|_{\mathbf{y}=0}$ of $\mathbf{y}$-free polynomials in $\mathcal{H}_\mu$. We can also describe the elements of $\mathcal{H}_\mu|_{\mathbf{y}=0}$ as linear combinations of polynomials of form $\partial\mathbf{x}^{\mathbf{a}}\partial\mathbf{y}^{\mathbf{b}}\Delta_\mu(\mathbf{x}, \mathbf{y})$ that contain none of the y_i. This forces $\mathbf{b}$ to be such that

$$\mathbf{y}^{\mathbf{b}} = \prod_{(i,j)\in\mu} y_{\tau(i,j)}^{i}$$

for some injective tableau τ of shape μ. But this monomial is invariant under column-fixing permutations of the entries of τ. It follows that $\partial\mathbf{y}^{\mathbf{b}}\Delta_\mu(\mathbf{x}, \mathbf{y})$ is nonzero and antisymmetric, when considered as a polynomial in the entries of a given column. It is thus divisible by $\Delta_\tau(\mathbf{x})$ and of the same degree, and this implies that $\partial\mathbf{y}^{\mathbf{b}}\Delta_\mu(\mathbf{x}, \mathbf{y}) = c\Delta_\tau(\mathbf{x})$ for some

---

[1] For a starting point on this subject, refer to [Manivel 01].

nonzero constant c. Moving on to the corresponding Frobenius characteristic, we get the specialization $\mathrm{Frob}_{qt}(\mathcal{H}_\mu)|_{q=0} = \mathrm{Frob}_t(\mathcal{G}_\mu)$. In turn, the resulting graded Frobenius characteristic may be written in terms of Kostka–Foulkes polynomials $K_{\lambda,\mu}(q)$ as

$$\mathrm{Frob}_q(\mathcal{G}_\mu) = \sum_\lambda q^{n(\mu)} K_{\lambda,\mu}(q^{-1}) s_\lambda.$$

Recall that

$$K_{\lambda,\mu}(q) = \sum_\tau q^{\mathrm{ch}(\tau)},$$

with sum over the set of semi-standard tableau τ of shape λ and content μ. (For the definition of the charge statistic, see Section 2.6.) For example, the $\mathbf{y}$-free part of the space $\mathcal{H}_{21}$ is readily seen to be

Basis element	bidegree	Frobenius characteristic
$x_1 - x_2, x_1 - x_3$	$(1,0)$	s_{21}
1	$(0,0)$	s_3

so that the resulting graded Frobenius characteristic is indeed $s_3 + ts_{21}$.

10.2   Generalization to Diagrams

We can extend the considerations of the previous section to analogous statements for n-cell diagrams. Recall from Section 1.3 that, given any n-cell diagram $\mathbf{d}$ with cells ordered in increasing lexicographic order, we define

$$\Delta_\mathbf{d}(\mathbf{x}, \mathbf{y}) := \det(x_k^j y_k^i)_{\substack{1 \leq k \leq n \\ (i,j) \in \mathbf{d}}}.$$

The vector space spanned by the partial derivatives of all orders in both sets of variables of a polynomial $f(\mathbf{x}, \mathbf{y})$ is denoted by $\mathcal{L}_\partial[f]$. In particular, we consider the case where $f(\mathbf{x}, \mathbf{y}) = \Delta_\mathbf{d}$, which is clearly antisymmetric under the diagonal action of $\mathfrak{S}_n$. Then the $\mathbf{d}$-*harmonic space* is the $\mathfrak{S}_n$-module $\mathcal{H}_\mathbf{d} = \mathcal{L}_\partial[\Delta_\mathbf{d}]$. Since $\Delta_\mathbf{d}$ is bihomogeneous, this module affords a natural bigrading. Denoting by $\pi_{r,s}[\mathcal{H}_\mathbf{d}]$ the subspace consisting of the bihomogeneous elements of degree r in $\mathbf{x}$ and degree s in $\mathbf{y}$, we have the direct sum decomposition

$$\mathcal{H}_\mathbf{d} = \bigoplus_{r,s \geq 0} \pi_{r,s}[\mathcal{H}_\mathbf{d}],$$

so that it makes sense to talk about the bigraded Hilbert series and bigraded Frobenius characteristic of $\mathcal{H}_\mathbf{d}$.

Many interesting properties of the space $\mathcal{H}_\mathbf{d}$ can be understood by studying the effect of the operator

$$p_{h,k}(\partial\mathbf{x}, \partial\mathbf{y}) := \sum_{i=1}^{n} \partial x_i^h \partial y_i^k$$

on the determinants $\Delta_\mathbf{d}(\mathbf{x}, \mathbf{y})$. The following proposition is similar to Proposition 8.9.

Proposition 10.1. *Let* $\mathbf{d}$ *be an* n*-cell diagram and* $h, k \geq 0$ *integers with* $h + k \geq 1$. *Let* $\mathbf{d}^{(i)}$ *be the diagram obtained from* $\mathbf{d}$ *by replacing the cell* (a_i, b_i) *with the cell* $(a_i - h, b_i - k)$. *Finally, define* $\epsilon(\mathbf{d}^{(i)})$ *to be 0 if one of the coordinates of* $(a_i - h, b_i - k)$ *is negative or if* $(a_i - h, b_i - k)$ *is some other cell of* $\mathbf{d}$, *and otherwise to be equal to the sign of the permutation that reorders* $\mathbf{d}$ *so that* (a_i, b_i) *takes the position that* $(a_i - h, b_i - k)$ *has in* $\epsilon(\mathbf{d}^{(i)})$ *in lexicographic order. Then*

$$p_{h,k}(\partial\mathbf{x}, \partial\mathbf{y})\Delta_\mathbf{d}(\mathbf{x}, \mathbf{y}) = \sum_{i=1}^{n} \epsilon(\mathbf{d}^{(i)}) \frac{a_i!}{(a_i - h)!} \frac{b_i!}{(b_i - k)!} \Delta_{\mathbf{d}^{(i)}}(\mathbf{x}, \mathbf{y}) \quad (10.3)$$

where $\mathbf{d}^{(i)}$ *is obtained from* $\mathbf{d}$ *by replacing the cell* (a_i, b_i) *with the cell* $(a_i - h, b_i - k)$.

For example, using the actual diagram $\mathbf{d}$ to denote the polynomial $\Delta_\mathbf{d}(\mathbf{x})$, we have

$$p_{11}(\partial\mathbf{x}, \partial\mathbf{y}) \left(\text{⌐} \right) = \text{⌐} \quad +2 \text{⌐} \quad -3 \text{⌐} \quad (10.4)$$

The four missing diagrams correspond either to cases where the translated cell has some negative coordinate or coincides with another cell. Observe that, applying Proposition 10.1 in the case of the determinant $\Delta_\mu(\mathbf{x}, \mathbf{y})$ for a partition μ, we find that for all h and k such that $h + k \geq 1$,

$$p_{h,k}(\partial\mathbf{x}, \partial\mathbf{y})\Delta_\mu(\mathbf{x}, \mathbf{y}) = 0, \quad (10.5)$$

since no cell can move without either "falling out" or overlapping an existing cell.

With the following generalization of Lemma 8.2 we can now give a more precise lower bound on the "multiplicity" of the regular representation in modules associated with general diagrams.

Lemma 10.2. *For any two n-cell diagram*

$$\mathbf{c} = \{(a_1, b_1), \ldots, (a_n, b_n)\} \quad and \quad \mathbf{d} = \{(p_1, q_1), \ldots, (p_n, q_n)\},$$

we have that $\Delta_{\mathbf{c}}(\partial \mathbf{x}, \partial \mathbf{y}) \Delta_{\mathbf{d}}(x; y) \neq 0$ *implies that there exists a permutation* σ *such that* $a_i \leq p_{\sigma(i)}$ *and* $b_i \leq q_{\sigma(i)}$ *for all* i.

For two n-cell diagrams such as in the lemma, we write $\mathbf{c} \preceq \mathbf{d}$ and define the polynomial

$$\tau_{\mathbf{d}}(q, t) := \sum_{\mathbf{c} \preceq \mathbf{d}} q^{n(\mathbf{c}')} t^{n(\mathbf{c})}.$$

Then, we have a "graded" lower bound[2] on the multiplicity of the trivial representation in $\mathcal{H}_{\mathbf{d}}$. All our partial results and experiments indicate that $\tau_{\mathbf{d}}(q, t)$ is in fact exactly the bigraded enumerator of the trivial representation in $\mathcal{H}_{\mathbf{d}}$, but we have no proof of this.

Remark 10.3. Observe that the upcoming proposition is only a statement about the multiplicity of the trivial representation in $\mathcal{H}_{\mathbf{d}}$ rather than one about the regular representation. However, recall that there is exactly one copy of the trivial representation in the regular representation. Hence, when $\mathcal{H}_{\mathbf{d}}$ is in fact a multiple of the regular representation, the two multiplicities (of the trivial and the regular) agree. However, $\mathcal{H}_{\mathbf{d}}$ is not always isomorphic to a direct sum of $\tau_{\mathbf{d}}(1, 1)$ copies of the left regular representation of $\mathfrak{S}_n$. For instance, this can be made apparent by an explicit computation for the diagram $\{(1, 0), (0, 1), (1, 2)\}$.

Proposition 10.4. *For any n-cell diagram* $\mathbf{d}$, *the bigraded enumerator of the trivial representation in* $\mathcal{H}_{\mathbf{d}}$ *is, coefficientwise, at least equal to the corresponding coefficient in* $\tau_{\mathbf{d}}(q, t)$.

To illustrate Proposition 10.4, consider the diagram $\mathbf{d} = \{(n - i - 1, i) \mid 0 \leq i \leq n - 1\}$. Then $\mathbf{c} \preceq \mathbf{d}$ if and only if there is an ordering $\{(a_1, b_1), \ldots, (a_n, b_n)\}$ of the cells of $\mathbf{c}$ such that $a_i \leq n - i - 1$, and $b_i \leq i$. One can show that in this case $\tau_{\mathbf{d}}(1, 1) = (n + 1)^{n-1}$, and even more precisely that $q^{\binom{n}{2}} \tau_{\mathbf{d}}(q, 1/q) = [n + 1]_q^{n-1}$.

---

[2]This is to say that dimensions are bounded componentwise. Thus, the actual Hilbert series of the trivial isotypic component is termwise bounded by the relevant polynomial.

10.3   Punctured Partition Diagrams

A nicely behaved class of examples is that of diagrams[3] $\mathbf{d} = \mu/i, j$, obtained by removing the cell (i, j) from the diagram of a partition μ of $n + 1$. We consider, for a fixed cell (i, j) the symmetric functions recursively defined by

$$H_{\mu/i,j} = \frac{t^\ell - q^{a+1}}{t^\ell - q^a} H_{\mu/i,j+1} + \frac{t^{\ell+1} - q^a}{t^\ell - q^a} H_{\mu/i+1,j}$$
$$- \frac{t^{\ell+1} - q^{a+1}}{t^\ell - q^a} H_{\mu/i+1,j+1}, \qquad (10.6)$$

with a and ℓ respectively denoting the arm and leg length of the cell (i, j) in μ, with the following boundary conditions:

(1) $H_{\mu/i,j} = 0$ whenever (i, j) is not in μ,

(2) $H_{\mu/i,j} = H_\nu$ when (i, j) is a corner cell of μ such that $\mu/i, j$ is the partition ν.

A priori, the solution $H_{\mu/i,j}$ of recurrence (10.6) is a $\mathbb{Q}(q, t)$-linear combination of the H_ν, with ν running over partitions that can be obtained from μ by removing a corner that sits in the shadow of the cell (i, j). (Recall from Section 1.4 that the shadow of (i, j) is the set of cells (s, t) such that $s \geq i$ and $t \geq j$.) Let us illustrate with $\mu = 32$. The corners of 32 are the cells $(2, 0)$ and $(1, 1)$, and we have the boundary values

$$H_{32/1,1} = H_{31} = s_4 + (q + t + q^2)s_{31} + (q^2 + qt)s_{22}$$
$$+ (qt + q^2t + q^3)s_{211} + q^3t s_{1111},$$
$$H_{32/2,0} = H_{22} = s_4 + (q + t + qt)s_{31} + (q^2 + t^2)s_{22}$$
$$+ (qt + q^2t + qt^2)s_{211} + q^2t^2 s_{1111}.$$

Applying recurrence (10.6), we successively find

$$H_{32/0,1} = (q + 1)H_{31},$$
$$[29pt]H_{32/1,0} = \frac{q - t^2}{q - t} H_{31} + \frac{q^2 - t}{q - t} H_{22}, \qquad (10.7)$$
$$H_{32/0,0} = \frac{(q + 1)(q - t^2)}{q - t} H_{31} + \frac{q^3 - t}{q - t} H_{22}. \qquad (10.8)$$

---

[3] We simplify the notation $\mu \setminus \{(i, j)\}$ to $\mu/i, j$.

It may come as a surprise that the right-hand sides of the last two equalities are in fact Schur-positive. Namely, we have:

$$
\begin{aligned}
H_{32/1,0} = {} & (q + t + 1)s_4 + (q^2 t + 2q^2 + 3qt + t^2 + q + t)s_{31} \\
& + (q^3 + q^2 t + qt^2 + q^2 + qt + t^2)s_{22} \\
& + (q^3 t + q^2 t^2 + q^3 + 3q^2 t + 2qt^2 + qt)s_{211} \\
& + q^2 t(qt + q + t)s_{1111}, \\
H_{32/0,0} = {} & (q^2 + qt + q + t + 1)s_4 \\
& + (q^3 t + 2q^3 + 3q^2 t + qt^2 + 2q^2 + 3qt + t^2 + q + t)s_{31} \\
& + (q^4 + q^3 t + q^2 t^2 + q^3 + 2q^2 t + qt^2 + q^2 + qt + t^2)s_{22} \\
& + (q^4 t + q^3 t^2 + q^4 + 3q^3 t + 2q^2 t^2 + q^3 + 3q^2 t + 2qt^2 + qt)s_{211} \\
& + (q^4 t^2 + q^4 t + q^3 t^2 + q^3 t + q^2 t^2)s_{1111}.
\end{aligned}
$$

This fact would clearly follow from the following conjecture, which first appeared in [Bergeron et al. 99b].

Conjecture 10.5 *For all partitions μ and all cells (i,j) of μ, the bigraded Frobenius characteristic of the space $\mathcal{H}_{\mu/i,j}$ is given by the symmetric function $H_{\mu/i,j}$.*

Proposition 10.6 below shows that this conjecture actually holds for the case $(i,j) = (0,0)$. Recall that $s_1^\perp$ denotes the operator that is adjoint (for the usual symmetric function scalar product) to multiplication by s_1. Then a direct reformulation of a Pieri rule[4] for the Macdonald functions P_μ of Section 9.1 in terms of the H_μ states that

$$
s_1^\perp H_\mu = \sum_{\nu \to \mu} c_{\mu\nu} H_\nu, \tag{10.9}
$$

with $c_{\mu\nu} = c_{\mu\nu}(q,t)$ lying in the ring of rational fractions in q and t. Recall that we write $\nu \to \mu$ when μ covers ν in Young lattice (see Section 1.7), so that ν is obtained by removing a corner cell from μ. The existence of an expansion of $s_1^\perp H_\mu$ in terms of the integral form Macdonald functions is guaranteed by the fact that the H_μ form a basis, but the point here is that this expansion involves only the H_ν for ν preceding μ in the Young lattice.

As a matter of fact, the right-hand side of (10.8) is precisely the expression that one would find for $s_1^\perp H_{32}$ using Macdonald explicit formulas reformulated in the form (10.9). See Appendix A for an explicit description.

There is indeed a close connection between Conjecture 10.5 and (10.9) which goes as follows. Let ρ denote the partition corresponding to the

---

[4]See [Macdonald 95, Chapter VI, Section 6].

Figure 10.1. Shadow of (i,j) in μ. The partition $\mu - \rho + \tau$.

shadow of (i,j) in μ, and let "$\mu - \rho + \tau$" stand for the diagram obtained by replacing ρ by τ inside μ (thus deleting a cell) as illustrated in Figure 10.1. This is the partition obtained by removing from μ the cell by which ρ and τ differ. As shown in [Bergeron et al. 99b], recurrence (10.6) is equivalent to the statement that

$$H_{\mu/i,j} = \sum_{\tau \to \rho} c_{\rho\tau} H_{\mu - \rho + \tau},$$

for each choice of (i,j) in μ.

Our upcoming Proposition 10.6 relies on a better understanding of the module $\mathcal{H}_{\mu/0,0}$. Let $\mathbf{x}' = x_1, \ldots, x_{n+1}$ and $\mathbf{y}' = y_1, \ldots, y_{n+1}$, and consider the bigrading preserving linear transformation

$$\Psi := \mathbb{C}[\mathbf{x}', \mathbf{y}'] \longrightarrow \mathbb{C}[\mathbf{x}, \mathbf{y}],$$

where $\Psi\big(f(\mathbf{x}', \mathbf{y}')\big)$ obtained by setting $x_{n+1} = y_{n+1} = 0$ in $f(\mathbf{x}', \mathbf{y}')$. For μ a partition of $n+1$, it is easily verified that $\Psi\big(\Delta_\mu(\mathbf{x}')\big) = \Delta_{\mu/0,0}(\mathbf{x})$. With a little more work it follows that $\Psi(\mathcal{H}_\mu) = \mathcal{H}_{\mu/0,0}$ and that Ψ restricts to a $\mathfrak{S}_n$-module isomorphism

$$\Psi \colon \mathcal{H}_\mu|_{\mathfrak{S}_n} \longrightarrow \mathcal{H}_{\mu/0,0},$$

considering $\mathfrak{S}_n$ in $\mathfrak{S}_{n+1}$ as the subgroup of permutations fixing $n+1$. As previously mentioned, we have $\mathrm{Frob}_{q,t}(\mathcal{H}_\mu|_{\mathfrak{S}_n}) = s_1^\perp \, \mathrm{Frob}_{q,t}(\mathcal{H}_\mu)$. Thus, we get the following proposition which gives further credence to our conjecture. (More details can be found in [Bergeron et al. 99b].)

Proposition 10.6. *For all μ, $\mathrm{Frob}_{qt}(\mathcal{H}_{\mu/0,0}) = s_1^\perp \, \mathrm{Frob}_{qt}(\mathcal{H}_\mu)$. In particular, $\mathcal{H}_{\mu/0,0}$ is the restriction to $\mathfrak{S}_n$ of the $\mathfrak{S}_{n+1}$-representation $\mathcal{H}_\mu$, and we have*

$$\mathrm{Frob}_{qt}(\mathcal{H}_{\mu/0,0}) = \sum_{\nu \to \mu} c_{\mu\nu} H_\nu.$$

Let us now consider the specialization of at $q = t = 1$ of recurrence (10.6). When $a > 0$ we may set $t = 1$, multiply both sides by $(1-q^a)/(1-q)$, and then take the limit for q going to 1, to deduce that

$$aH_{\mu/i,j}(\mathbf{z};1,1) = (a+1)H_{\mu/i,j+1}(\mathbf{z};1,1) + aH_{\mu/i+1,j}(\mathbf{z};1,1)$$
$$- (a+1)H_{\mu/i+1,j+1}(\mathbf{z};1,1).$$

Similarly, when $\ell > 0$ we have

$$\ell H_{\mu/i,j}(\mathbf{z};1,1) = \ell H_{\mu/i,j+1}(\mathbf{z};1,1) + (\ell+1)H_{\mu/i+1,j}(\mathbf{z};1,1)$$
$$- (\ell+1)H_{\mu/i+1,j+1}(\mathbf{z};1,1).$$

Observe that boundary condition (2) becomes $H_{\mu/i,j} = h_1^n$ when (i,j) is a corner cell of μ such that $\mu/i,j$ is the partition ν. It follows that the solution of the recurrence is just a multiple of h_1^n. (Recall that h_1^n is the character of the regular representation.) It is easily checked that this multiple is exactly the number of cells that appear in the shadow of the cell (i,j) in μ. We denote this number by $w_{ij}(\mu)$. To summarize, we have

$$\mathrm{Frob}_{qt}(\mathcal{H}_{\mu/i,j})\,|_{q=t=1} = w_{ij}(\mu)h_1^n. \tag{10.10}$$

Exercise. Show that the coefficient of s_n in the solution $H_{\mu/i,j}$ of (10.3) is $B_\rho(q,t)$ (see Section 1.3), where ρ is the partition that corresponds to the shadow of (i,j) in μ. Also, give an expression for the coefficient of s_{1^n} in $H_{\mu/i,j}$.

Among the many other appealing and interesting aspects of Conjecture 10.5, the most striking is certainly the fact that its veracity would lead to a direct recursive proof of the $n!$-Theorem. Indeed, Proposition 10.6, together with (10.10), says that the space $\mathcal{H}_{\mu/0,0}$ decomposes into $n+1$ copies of the regular representation. But we already know that this should be the case if $\mathcal{H}_\mu$ is to be isomorphic to the regular representation of $\mathfrak{S}_{n+1}$. Indeed, in light of (5.15) we see that the restriction to $\mathfrak{S}_n$ of the regular representation of $\mathfrak{S}_{n+1}$ has Frobenius characteristic equal to $(n+1)h_1^n$.

The plausibility of Conjecture 10.5 is further underlined by the fact that the set

$$\{\Delta_{\mu/a,b}(\mathbf{x},\mathbf{y}) \mid (a,b) \in \mu, a \geq i, b \geq j\}$$

forms a basis of the set of antisymmetric polynomials of the space $\mathcal{H}_{\mu/i,j}$. Clearly, in view of Proposition 10.1, all elements of this set can be obtained from $\Delta_{\mu/i,j}$ (up to a nonzero constant c depending on (i,j) and (a,b)) as

$$p_{h,k}(\partial\mathbf{x},\partial\mathbf{y})\Delta_{\mu/i,j}(\mathbf{x},\mathbf{y}) = c\Delta_{\mu/a,b}(\mathbf{x},\mathbf{y})$$

by choosing $h = a-i$ and $k = b-j$. In particular, this implies that $\mathcal{H}_{\mu/a,b}$ is a subspace of $\mathcal{H}_{\mu/i,j}$.

Most of the considerations of this section have been extended to k-punctured diagrams. A brief account of this extension can be found in [Aval 02]. Finally, let us mention that all consequences of Conjecture 10.5 that result from setting all of the variables y_i equal to 0 have been shown to be true in [Aval et al. 02].

10.4   Intersections of Garsia–Haiman Modules

The so-called SF-heuristic (see [Bergeron and Garsia 99]), gives a decomposition of the Garsia–Haiman modules $\mathcal{H}_\nu$ that goes hand in hand with the study of the intersection of such modules for partitions that are covered by a given partition in the Young lattice. One of the interesting features of this decomposition is that it allows a parallel decomposition of the associated integral form Macdonald functions. Moreover, its sheds surprising light on the study of the symmetric functions $H_{\mu/i,j}$ that we have just associated with punctured diagrams. We now briefly outline this still conjectural decomposition, referring the reader to [Bergeron and Garsia 99] for further details. To make some formulas in this section more readable, for each diagram $\mathbf{d}$ we define $T_{\mathbf{d}} := q^{n(\mathbf{d}')} t^{n(\mathbf{d})}$.

For a fixed partition μ of $n+1$, let $\pi(\mu)$ be the set of predecessors of μ in the Young lattice, i.e., the partitions obtained by removing one of the corners of μ. Recall (from Section 1.7) that for each element $\nu \in \pi(\mu)$ we write $\nu \to \mu$ and we denote the relevant corner by μ/ν. For any subset A of $\pi(\mu)$ of cardinality $a > 0$, we wish to study the space $\mathcal{H}_A := \bigcap_{\alpha \in A} \mathcal{H}_\alpha$. A striking experimentally observed property of the module $\mathcal{H}_A$ is that it appears to always have dimension $n!/a$. This is made much clearer using the SF-heuristic. The first assertion of the SF-heuristic is that in the linear span $\mathcal{L}[H_\nu \mid \nu \in A]$ there is a Schur-positive symmetric function $\varphi_A(\mathbf{z})$ given by the formula

$$\varphi_A(\mathbf{z}) = \sum_{\alpha \in A} \Big(\prod_{\beta \in A/\{\alpha\}} \frac{-T_\beta}{T_\alpha} \Big) H_\alpha(\mathbf{z}; q, t), \qquad (10.11)$$

which corresponds precisely to the Frobenius characteristic of the $\mathfrak{S}_n$-module $\mathcal{H}_A$. If $\nu \in A$, then the SF-heuristic also asserts that there is a decomposition of the integral form Macdonald function H_ν of the form

$$H_\nu(\mathbf{z}; q, t) = \prod_{\alpha \in A \setminus \{\nu\}} \Big(1 - \frac{\nabla}{T_\alpha} \Big) \varphi_A(\mathbf{z}). \qquad (10.12)$$

In fact, there is a proposed explanation in [Bergeron and Garsia 99] of this identity in terms of a decomposition of the module $\mathcal{H}_\nu$ as direct sum of specific subspaces. Finally, if we let $\varphi_A^{(k)}(\mathbf{z}) := (-\nabla)^{a-k}\varphi_A(\mathbf{z})$ for $1 \leq k \leq a - 1$, the SF-heuristic states that for all cardinality-k subsets B of A, the coefficient of any Schur function in the Schur basis expansion of the symmetric function

$$\varphi_A^B(\mathbf{z}) := \prod_{\beta \in A \backslash B} \frac{1}{T_\beta} \varphi_A^{(k)}(\mathbf{z}) \tag{10.13}$$

is a positive integer polynomial in q and t. Thus, (10.12) gives an expansion of H_ν as a sum of 2^{a-1} Schur-positive function:

$$H_\nu(\mathbf{z}; q, t) = \sum_{\nu \in B \subseteq A} \varphi_A^B(\mathbf{z}). \tag{10.14}$$

For example, the integral form Macdonald function

$$\begin{aligned}
H_{32} = {}&s_5 + (q^2 + qt + q + t)s_{41} + (q^3 + q^2t + q^2 + qt + t^2)s_{32} \\
&+ (q^3t + q^3 + 2q^2t + qt^2 + qt)s_{311} \\
&+ (q^4 + q^3t + q^2t^2 + q^2t + qt^2)s_{221} \\
&+ (q^4t + q^3t^2 + q^3t + q^2t^2)s_{2111} + q^4t^2 s_{11111}
\end{aligned}$$

decomposes this way into the sum $\varphi_0 + \varphi_1/(q^2t^4) + \varphi_1/(q^3t^3) + \varphi_2/(q^5t^7)$, with

$$\varphi_0 = s_5 + (q + t)s_{41} + (q^2 + qt + t^2)s_{32} + qts_{311} + (q^2t + qt^2)s_{221},$$

$$\varphi_1 = q^4t^4\big(s_{4,1} + (q + t)s_{311} + qts_{2111}\big),$$

$$\varphi_2 = q^7t^7\big((q + t)s_{32} + qts_{311} + (qt + t^2 + q^2)s_{221}$$
$$+ qt(q + t)s_{2111} + q^2t^2 s_{11111}\big).$$

But the surprise here is that there are similar expressions for H_{311} and H_{221} in terms of the same φ_0, φ_1 and φ_2.

Some headway has been made towards understanding the SF-heuristic in terms of the modules involved, see [Bergeron and Hamel 00]. It is also noteworthy that all the assertions of the SF-heuristic can be extended to the context of punctured diagrams, provided that the point removed is the same for each diagram considered. Moreover, if this point is the origin $(0, 0)$ then the punctured version follows from the original SF-heuristic.

10.5   Diagonal Harmonics

As a direct consequence of Proposition 10.1, every space $\mathcal{H}_\mu$ with μ a partition of n is a submodule of the larger space of "diagonal harmonic

polynomials". This immediately follows from (10.5), which implies that
any element $f(\mathbf{x}, \mathbf{y})$ of $\mathcal{H}_\mu$ is killed by diagonally symmetric operators, since
we must have $p_{h,k}(\partial\mathbf{x}, \partial\mathbf{y})f(\mathbf{x}, \mathbf{y}) = 0$. Our new story here starts with the
ring $\mathcal{R}^{\mathfrak{S}_n}$ of diagonally symmetric polynomials, namely those polynomials
in $\mathcal{R} = \mathbb{C}[\mathbf{x}, \mathbf{y}]$ such that $\sigma \cdot f(\mathbf{x}, \mathbf{y}) = f(\mathbf{x}, \mathbf{y})$ for all permutations σ.
One easily shows that this ring in generated by the *diagonal power sum
polynomials* $p_{h,k}(\mathbf{x}, \mathbf{y}) := \sum_{i=1}^n x_i^h y_i^k$. Then, in the spirit of what was done
in earlier sections, we consider the diagonal-action-invariant scalar product
on $\mathcal{R}$,

$$\langle f(\mathbf{x}, \mathbf{y}), g(\mathbf{x}, \mathbf{y})\rangle := f(\partial\mathbf{x}, \partial\mathbf{y})g(\mathbf{x}, \mathbf{y})|_{\mathbf{x}=0, \mathbf{y}=0}, \qquad (10.15)$$

for which we can make observations similar to those of Section 8.2. We
can define (just as in the case of one set of variables) the $\mathfrak{S}_n$-module $\mathcal{D}_n$ of
diagonal harmonic polynomials to be the set of polynomials $f(\mathbf{x}, \mathbf{y})$ such
that $p_{h,k}(\partial\mathbf{x}, \partial\mathbf{y})f(\mathbf{x}, \mathbf{y}) = 0$ for all (h, k) such that $h + k > 0$. Thus, we
make sense of our opening remark, that for each partition μ, the space $\mathcal{H}_\mu$
is a submodule of $\mathcal{D}_n$.

For example, with $n = 2$, diagonal harmonic polynomials are the polynomial solutions of the system of differential equations

$$\frac{\partial}{\partial x_1}f(\mathbf{x}, \mathbf{y}) + \frac{\partial}{\partial x_2}f(\mathbf{x}, \mathbf{y}) = 0,$$

$$\frac{\partial}{\partial y_1}f(\mathbf{x}, \mathbf{y}) + \frac{\partial}{\partial y_2}f(\mathbf{x}, \mathbf{y}) = 0,$$

$$\frac{\partial^2}{\partial x_1^2}f(\mathbf{x}, \mathbf{y}) + \frac{\partial^2}{\partial x_2^2}f(\mathbf{x}, \mathbf{y}) = 0,$$

$$\frac{\partial^2}{\partial x_1\partial y_1}f(\mathbf{x}, \mathbf{y}) + \frac{\partial^2}{\partial x_2\partial y_2}f(\mathbf{x}, \mathbf{y}) = 0,$$

$$\frac{\partial^2}{\partial y_1^2}f(\mathbf{x}, \mathbf{y}) + \frac{\partial^2}{\partial y_2^2}f(\mathbf{x}, \mathbf{y}) = 0.$$

One readily checks that the associated solution set has basis $\{1, x_1 -
x_2, y_1 - y_2\}$. We see here that $\mathcal{D}_2$ indeed contains both Garsia–Haiman
modules $\mathcal{H}_2$ and $\mathcal{H}_{11}$ as submodules, since they are respectively spanned
by the sets $\{1, x_1 - x_2\}$ and $\{1, y_1 - y_2\}$. As we will see, the fact that they
sum up to $\mathcal{D}_2$ is entirely exceptional. Indeed, for $n > 2$, this is never the
case.

It is clear that we have the inclusion of $\mathcal{D}_n$ into the larger space $\mathcal{H}_{\mathfrak{S}_n \times \mathfrak{S}_n}$
of $\mathfrak{S}_n \times \mathfrak{S}_n$-harmonic polynomials, considered as a $\mathfrak{S}_n$-module for the diagonal action as described in Section 8.7. However, the story takes a radically
different turn here. After having explicitly computed the relevant spaces
for small values of n, Garsia and Haiman were led to conjecture that the
dimension of $\mathcal{D}_n$ is surprisingly nice from the combinatorial point of view.

Indeed, it appeared to be equal to $(n+1)^{n-1}$, which corresponds to the enumeration of several interesting combinatorial structures. In fact, since the space $\mathcal{D}_n$ is bihomogeneous and $\mathfrak{S}_n$-invariant, we can compute $\mathrm{Frob}_{qt}(\mathcal{D}_n)$, the associated bigraded Frobenius characteristic for which Haiman conjectured a rather complicated formula involving integral form Macdonald functions. Studying this formula, the author and A. Garsia were led to introduce the linear operator ∇, discussed in Section 9.6, in order to give a simpler description of $\mathrm{Frob}_{qt}(\mathcal{D}_n)$ as the effect of ∇ on the elementary symmetric function e_n. All of this is now a theorem that can be stated as follows.

Theorem 10.7 (Haiman 2003). *The bigraded Frobenius characteristic of the space of diagonal harmonics is given by the formula $\mathrm{Frob}_{qt}(\mathcal{D}_n) = \nabla(e_n)$. In particular, this implies that the space of diagonal harmonics has dimension*

$$\dim \mathcal{D}_n = (n+1)^{n-1}.$$

Operator Description

Another nice way of describing the space of diagonal harmonics involves the use of the operators

$$E_k := \sum_{i=1}^{n} y_i \partial x_i^k, \quad k \geq 1. \tag{10.16}$$

Since they commute among themselves, they generate under composition and addition a space spanned by the partition-indexed operators $E_\mu := E_{\mu_1} \cdots E_{\mu_r}$. If μ is a length-r partition of s and $f(\mathbf{x}, \mathbf{y})$ is a bihomogeneous polynomial of bidegree (m, k), we clearly have that $E_\mu f(\mathbf{x}, \mathbf{y})$ is of bidegree $(m - s, k + r)$. For this reason we say that the bidegree of E_μ is $(-s, r)$. Observe that $E_n \Delta_n(\mathbf{x}) = 0$ since all variables appear with a power at most $n - 1$ in Δ_n. Just as clearly, $E_\mu \Delta_n(\mathbf{x})$ is necessarily an antisymmetric polynomial, because E_μ is diagonally symmetric. The effect of the E_k on diagram determinants is especially easy to describe. For example, with the same conventions as in (10.4), we have

$$E_2 \left(\quad \right) = 6 \qquad + 2 \qquad - 2 \qquad \tag{10.17}$$

From the material developed by Haiman in his proof of Theorem 10.7 we get the following.

Theorem 10.8. *The subspace $\mathcal{A}_n \subset \mathcal{D}_n$ of antisymmetric polynomials is spanned by the polynomials $E_\mu \Delta_n(\mathbf{x})$, where μ runs over all partitions of $s \leq \binom{n}{2}$ with parts of size at most $n - 1$. Moreover, the bigraded Hilbert series of $\mathcal{A}_n$ is given by*

$$\mathrm{Hilb}_{qt}(\mathcal{A}_n) = \langle \nabla(e_n), e_n \rangle \tag{10.18}$$

and specializes to the nth Catalan number when $q = t = 1$.

q, t-Catalan Polynomials

In view of the latter part of Theorem 10.8, the polynomial $\mathrm{Hilb}_{qt}(\mathcal{A}_n)$ is called the q,t-*Catalan polynomial* and is denoted by $C_n(q,t)$. Garsia and Haglund have given this polynomial a combinatorial description [Garsia and Haglund 02]. As of this writing, there is no known description of an explicit basis of $\mathcal{A}_n$, which would preferably be bihomogeneous. In other words, such a description could take the form of a family $\sum_\mu a_\mu E_\mu \Delta_n$ with the bidegree distribution given by $C_n(q,t)$. For small values of n, the q,t-Catalan polynomials are

$$C_1(q,t) = 1, \quad C_2(q,t) = q + t, \quad C_3(q,t) = q^3 + q^2 t + q t^2 + t^3 + qt,$$
$$C_4(q,t) = q^6 + q^5 t + q^4 t^2 + q^3 t^3 + q^2 t^4 + q t^5 + t^6 + q^4 t$$
$$\quad + q^3 t^2 + q^2 t^3 + q t^4 + q^3 t + q^2 t^2 + q t^3,$$
$$C_5(q,t) = q^{10} + q^9 t + q^8 t^2 + q^7 t^3 + q^6 t^4 + q^5 t^5 + q^4 t^6 + q^3 t^7 + q^2 t^8 + q t^9$$
$$\quad + t^{10} + q^8 t + q^7 t^2 + q^6 t^3 + q^5 t^4 + q^4 t^5 + q^3 t^6 + q^2 t^7 + q t^8$$
$$\quad + q^7 t + 2 q^6 t^2 + 2 q^5 t^3 + 2 q^4 t^4 + 2 q^3 t^5 + 2 q^2 t^6 + q t^7 + q^6 t$$
$$\quad + q^5 t^2 + 2 q^4 t^3 + 2 q^3 t^4 + q^2 t^5 + q t^6 + q^4 t^2 + q^3 t^3 + q^2 t^4.$$

To make the symmetries of these polynomials more apparent, we present them in matrix notation. We introduce a matrix C_n, setting

$$C_n(i,j) = \text{coefficient of } t^{k-i+1} q^{j-1} \text{ in } C_n(q,t) \tag{10.19}$$

with k equal to the t-degree of $C_n(q,t)$. Thus $C_3(q,t)$ becomes

$$C_3 = \begin{pmatrix} 1 & & & \\ 0 & 1 & & \\ 0 & 1 & 1 & \\ 0 & 0 & 0 & 1 \end{pmatrix},$$

where, to simplify even further, we have omitted some of the 0 entries. Observe the symmetries in the matrix

$$
C_5 = \begin{bmatrix}
1 & & & & & & & & & & \\
0 & 1 & & & & & & & & & \\
0 & 1 & & & & & & & & & \\
0 & 1 & 1 & 1 & & & & & & & \\
0 & 1 & 2 & 1 & 1 & & & & & & \\
0 & 0 & 1 & 2 & 1 & 1 & & & & & \\
0 & 0 & 1 & 2 & 2 & 1 & 1 & & & & \\
0 & 0 & 0 & 1 & 2 & 2 & 1 & 1 & & & \\
0 & 0 & 0 & 0 & 1 & 1 & 2 & 1 & 1 & & \\
0 & 0 & 0 & 0 & 0 & 0 & 1 & 1 & 1 & 1 & \\
0 & 0 & 0 & 0 & 0 & 0 & 0 & 0 & 0 & 0 & 1
\end{bmatrix}.
$$

Exercise. An *inner corner* (i,j) of C_n is a nonzero entry of C_n such that $C_n(k,\ell) = 0$ if $k < i$ or $j < \ell$. Show that the sum of the inner corner entries of C_n is at least equal to the number of partitions of n.

Let us add in conclusion that all of $\mathcal{D}_n$ can be obtained by derivation of elements of $\mathcal{A}_n$. More precisely, $\mathcal{D}_n$ is the smallest vector space containing the Vandermonde determinant Δ_n that is closed under application of the operators E_k, for $1 \leq k \leq n-1$, and under partial derivatives. We write this as

$$
\mathcal{D}_n = \mathcal{L}_{\partial,E}[\Delta_n].
$$

In other words, $\mathcal{D}_n$ is spanned by the set

$$
\{\partial \mathbf{x}^{\mathbf{a}} \partial \mathbf{y}^{\mathbf{b}} E_\mu \Delta_n(\mathbf{x}) \mid \mathbf{a}, \mathbf{b} \in \mathbb{N}^n \quad \text{and} \quad \mu \vdash k \in \mathbb{N}\}. \tag{10.20}
$$

10.6   Specialization of Parameters

Before coming to the core of this section, let us introduce some facts about q-Lagrange inversion as defined in [Garsia 81].

q-Lagrange Inversion

By definition, the *(right) q-inverse* of a series $F(z) = \sum_{k \geq 1} f_k z^k$, with $f_k \in \mathbb{C}[e_1, e_2, \ldots](q)$ and $f_1 = 1$, is the series $G(z) = \sum_{k \geq 1} g_k z^k$ such that

$$
\sum_{k \geq 1} f_k G(z) G(qz) \cdots G(q^{k-1} z) = z. \tag{10.21}
$$

We are using here a notion of q-composition of series that is not associative. Hence, the notions of right-inverse and left-inverse are not equivalent.

However, $F(z)$ is indeed the *left q^{-1}-inverse* of its right inverse $G(z)$, which is to say that

$$\sum_{k \geq 1} g_k F(z) F(z/q) \cdots F(z/q^{k-1}) = z. \qquad (10.22)$$

We are going to compute the q-inverse of the series

$$F(z) = z \left(\sum_{n \geq 0} e_n z^n \right)^{-1}. \qquad (10.23)$$

Just as in Section 6.5, we can think of the solution as being a "generic" q-inverse, since elementary symmetric functions are algebraically independent, and so they can be specialized to the coefficients of a given series. In fact, as shown in [Garsia and Haiman 96a], the right q-inverse of the series (10.23) is

$$G(z) = \sum_{n \geq 0} q^n g_n(q) z^n,$$

with

$$g_n(q) = \sum_{\mu \vdash n} q^{n(\mu')} h_\mu \left[\frac{\mathbf{x}}{1 - q} \right] f_\mu [1 - q]. \qquad (10.24)$$

First Specialization, $t = 1$

Setting $t = 1$ in formulas involving integral form Macdonald functions $H_\mu(\mathbf{z}; q, t)$ is particularly interesting, especially for the effect on the operator ∇. Recall from (9.13) that

$$H_\mu(\mathbf{z}; q, 1) = H_{\mu_1}(\mathbf{z}; q, 1) \cdots H_{\mu_k}(\mathbf{z}; q, 1).$$

Since $n(\mu') = \sum_i \binom{\mu_i}{2}$, we get, after a computation, that

$$\nabla_{t=1}(H_\mu) = \nabla_{t=1}(H_{\mu_1}) \cdots \nabla_{t=1}(H_{\mu_k}),$$

so the specialization $t = 1$ transforms ∇ into a multiplicative[5] linear operator. In particular, we deduce from this multiplicativity of $\nabla_{t=1}$ and the Jacobi–Trudi identity that

$$\nabla_{t=1}(s_\lambda) = \det\left(\nabla_{t=1}(e_{\lambda'_i + j - i}) \right). \qquad (10.25)$$

For example, setting $t = 1$ in previously computed values of ∇, we get

$$\nabla(e_1) = s_1, \quad \nabla(e_2) = s_2 + (q+1)s_{11},$$
$$\nabla(e_3) = s_3 + (q^2 + 2q + 2)s_{21} + (q^3 + q^2 + 2q + 1)s_{111},$$

---

[5]This is certainly not the case for ∇.

and (10.25) gives

$$
\begin{aligned}
\nabla_{t=1}(s_{21}) &= \det \begin{pmatrix} \nabla_{t=1}(e_2) & \nabla_{t=1}(e_3) \\ 1 & \nabla_{t=1}(e_1) \end{pmatrix} \\
&= s_1\big(s_2 + (q+1)s_{11}\big) - \big(s_3 + (q^2 + 2q + 2) \\
&\qquad s_{21} + (q^3 + q^2 + 2q + 1)s_{111}\big) \\
&= -q(q+)s_{21} - q(q^2 + q + 1)s_{111},
\end{aligned}
$$

which agrees with values computed to illustrate Conjecture 9.1. This multiplicativity of $\nabla_{t=1}$ is exploited by C. Lenart in his proof (see [Lenart 00]) of the specialization at $t = 1$ of Conjecture 9.1.

For our next development, we have from (9.13) that

$$
H_{(n)}(\mathbf{z}; q, 1) = \prod_{k=1}^{n} (1 - q^k) h_n \Big[\frac{\mathbf{z}}{1 - q} \Big],
$$

so we can reformulate the definition of $\nabla_{t=1}$ in the form

$$
\nabla_{t=1} h_\mu \Big[\frac{\mathbf{z}}{1 - q} \Big] = q^{n(\mu')} h_\mu \Big[\frac{\mathbf{z}}{1 - q} \Big]. \tag{10.26}
$$

Let us then start with the dual Cauchy formula

$$
e_n(\mathbf{xy}) = \sum_{\mu \vdash n} h_\mu(\mathbf{x}) f_\mu(\mathbf{y}),
$$

with $f_\mu = \omega m_\mu$ the "forgotten" basis. Now, substitute $\mathbf{z}/(1 - q)$ for $\mathbf{x}$ and $1 - q$ for $\mathbf{y}$ to get the expansion

$$
e_n(\mathbf{z}) = \sum_{\mu \vdash n} h_\mu \Big[\frac{\mathbf{z}}{1 - q} \Big] f_\mu[1 - q] \tag{10.27}
$$

of $e_n(\mathbf{z})$ in the basis $\{ h_\mu[\mathbf{z}/(1 - q)] \mid \mu \vdash n \}$. This makes it easy to compute that

$$
\nabla_{t=1}(e_n) = \sum_{\mu \vdash n} q^{n(\mu')} h_\mu \Big[\frac{\mathbf{z}}{1 - q} \Big] f_\mu[1 - q]. \tag{10.28}
$$

We just encountered this expression (up to a factor q^n) in (10.24) as the coefficient of z^n in the q-Lagrange inverse of the series

$$
\sum_{k \geq 0} (-1)^k h_k z^{k+1} = z \Big(\sum_{n \geq 0} e_n z^n \Big)^{-1}.
$$

Among the manifold implications of this observation, we find that the polynomials $\langle \nabla_{t=1}(e_n), e_n \rangle$ are none other than our previously encountered

q-analogs $\mathcal{C}_n(q)$ of the Catalan numbers, namely those that satisfy the q-recurrence

$$\mathcal{C}_n(q) = \sum_{k=0}^{n-1} q^k \mathcal{C}_k(q) \mathcal{C}_{n-1-k}(q). \tag{10.29}$$

To see this, we consider the generating series

$$\mathcal{C}(z; q) := \sum_{k \geq 0} q^{k-1} \mathcal{C}_k(q) z^{k+1}$$

and check that it satisfies the q-algebraic equation

$$\mathcal{C}(z; q) = 1 + \mathcal{C}(z; q)\mathcal{C}(qz; q).$$

Second Specialization, $t = 1/q$

We now consider another interesting specialization: $t = 1/q$. From (9.15) we may directly check that $s_\mu[\mathbf{x}/(1-q)]$ is an eigenfunction of $\nabla_{t=1/q}$ with eigenvalue $q^{n(\mu')-n(\mu)}$. This, together with the dual Cauchy formula gives

$$q^{\binom{n}{2}} \nabla_{t=1/q}(e_n) = \sum_{\mu \vdash n} q^{\binom{n}{2}+n(\mu')-n(\mu)} s_\mu\left[\frac{\mathbf{x}}{1-q}\right] s_{\mu'}[1-q]. \tag{10.30}$$

On the other hand, we have already shown (see (4.26)) that $s_\mu[1-q] = 0$, except when μ is a hook. It follows that the above sum only involves partitions of the form $(n-k, 1^k)$ with $1 \leq k \leq n-1$. For these we have the evaluation

$$q^{\binom{n}{2}+n(\mu')-n(\mu)} s_{\mu'}[1-q] = (-1)^k q^{nk+k}(1-q) = s_{\mu'}[1-q^{n+1}]\frac{1-q}{1-q^{n+1}}.$$

Another use of the Cauchy formula gives

$$q^{\binom{n}{2}} \nabla_{t=1/q}(e_n) = \frac{1}{[n+1]_q} e_n(\mathbf{x}[n+1]_q). \tag{10.31}$$

Now, to compute $\langle q^{\binom{n}{2}} \nabla_{t=1/q}(e_n), e_n \rangle$ we observe that

$$e_n(\mathbf{x}[n+1]_q) = \sum_{\lambda \vdash n} s_\lambda(\mathbf{x}) s_{\lambda'}([n+1]_q).$$

Thus,

$$\langle q^{\binom{n}{2}} \nabla_{t=1/q}(e_n), e_n \rangle = \frac{1}{[n+1]_q} h_n([n+1]_q). \tag{10.32}$$

But we also have

$$\sum_{k \geq 0} h_k([n+1]_q) z^k = \exp\left(\sum_{j \geq 1} p_j [1 + q + \cdots + q^n] z^j / j\right)$$

$$= \exp\left(\sum_{j \geq 1} (1 + q^j + \cdots + q^{nj}) z^j / j\right)$$

$$= \prod_{s=0}^{n} \frac{1}{1 - q^s z}.$$

It follows that

$$h_k([n+1]_q) = \begin{bmatrix} n+k \\ k \end{bmatrix}_q.$$

Substituting in (10.32), we find

$$\langle \nabla_{t=1/q}(e_n), e_n \rangle = q^{-\binom{n}{2}} \frac{1}{[n+1]_q} \begin{bmatrix} 2n \\ n \end{bmatrix}_q. \tag{10.33}$$

We now compute the coefficient of $p_1^n/n!$ in $1/[n+1]_q e_n(\mathbf{x}[n+1]_q)$, substituting $\mathbf{y} = [n+1]_q$ in the formula $e_n(\mathbf{xy}) = \sum_\lambda \frac{1}{z_\lambda} p_\lambda(\mathbf{x}) p_\lambda(\mathbf{y})$, to get

$$\frac{1}{[n+1]_q} e_n(\mathbf{x}[n+1]_q) = \frac{1}{[n+1]_q} \sum_\lambda \frac{1}{z_\lambda} p_\lambda(\mathbf{x}) p_\lambda([n+1]_q). \tag{10.34}$$

The coefficient of $p_1^n/n!$ in the right-hand side is clearly $[n+1]_q^{n-1}$.

Main Specialization Results

We now make use of the fact that the complete homogeneous symmetric functions h_n are algebraically independent, which allows us to specialize each h_n in any manner we desire. For our purpose, let us set

$$h_k \longmapsto \begin{cases} 1 & \text{if } k = 1, \\ 0 & \text{if } k \geq 2, \end{cases}$$

so that $\sum_{n \geq 0} h_n(-z)^n$ becomes $1 - z$. This implies that all of the elementary symmetric functions become equal to 1, since we have the evident identity (see (3.16))

$$1 = \left(\sum_{n \geq 0} e_n z^n\right) \left(\sum_{n \geq 0} h_n(-z)^n\right)$$

$$= \left(\sum_{n \geq 0} e_n z^n\right)(1 - z),$$

and thus $\sum_{n\geq 0} e_n z^n = (1 - z)^{-1}$. Using the Jacobi–Trudi formula, we can then check that s_λ maps to $\langle s_\lambda, e_n \rangle$ under the specialization of the h_k considered. Thus, we can now verify that

$$\sum_{n\geq 0} q^{n-1} \langle \nabla_{t=1}(e_n), e_n \rangle z^{n+1}$$

is indeed the q-Lagrange inverse of $\sum_k (-1)^k \langle h_k, e_k \rangle z^{k+1} = z - z^2$. Moreover, if we specialize the resulting polynomials $\langle \nabla_{t=1}(e_n), e_n \rangle$ at $q = 1$, we do get the Catalan numbers. To summarize:

Proposition 10.9. *Haiman's theorem, stating that* $\mathrm{Frob}_{qt}(\mathcal{D}_n) = \nabla(e_n)$, *implies that the bigraded series* $\mathrm{Hilb}_{qt}(\mathcal{H}_n)$ *is a q,t-analog of $(n+1)^{n-1}$. Moreover,*

$$q^{\binom{n}{2}} \mathrm{Hilb}_{qt}(\mathcal{D}_n)|_{t=1/q} = [n+1]_q^{n-1}. \tag{10.35}$$

It also implies that the bigraded multiplicity of the sign representation in the space $\mathcal{D}_n$ is the polynomial $\mathcal{C}_n(q,t) := \langle \nabla(e_n), e_n \rangle$, which is a q,t-analogue of the Catalan numbers. Moreover, we have $\mathcal{C}_n(q,1) = \mathcal{C}_n(q)$, and $q^{\binom{n}{2}} \mathcal{C}_n(q, 1/q) = \mathbf{C}_n(q)$.

To be clear about notational conventions regarding the Catalan q-analogs, we should consult Section 1.7. Observe that setting $t = 1/q$ corresponds to summing entries along diagonals of the q,t-Catalan matrices, while setting $t = 1$ corresponds to summing along columns in the q,t-Catalan matrices.

Combinatorial Descriptions

The recent past has seen a lot of exciting activity concerning possible combinatorial descriptions of the bigraded Frobenius characteristic $\nabla(e_n)$ of the space $\mathcal{D}_n$. In particular, we find in [Haglund et al. 05b] a beautiful conjecture for the expansion of $\nabla(e_n)$ in terms of the monomial basis. Before we can state it, we need to introduce the notion of a "d-inversion". Let $\lambda \subset \delta_n$ be a partition and τ a semi-standard tableau of shape $(\lambda + 1^n)/\lambda$ (see Figure 10.2). For such a tableau, let $\tau(i,j) = a$ and $\tau(k,\ell) = b$ with $a > b$. One says that the pair (a, b) forms a d-inversion if either

(1) $k + \ell = i + j$ and $i < k$

or

(2) $k + \ell = i + j - 1$ and $i > k$.

To better understand the conditions for a d-inversion, observe that for a cell (i, j) the value $i + j$ parametrizes the diagonal on which the cell sits.

Figure 10.2. Semi-standard tableau of shape $(42 + 1^5)/42$.

In the first case, we ask that both entries a and b appear on the same diagonal with a to the left of b, and in the second that they appear in consecutive diagonals with b to the left of a. We write $\text{dinv}(\tau)$ for the number of d-inversions of the tableau τ. Then the conjecture is that

$$\text{Frob}_{qt}(\mathcal{D}_n)(\mathbf{z}) = \sum_{\lambda \subseteq \delta_n} \sum_{\tau} t^{|\delta_n/\lambda|} q^{\text{dinv}(\tau)} \mathbf{z}_\tau, \tag{10.36}$$

where τ varies in the set of semi-standard Young tableaux of shape $(\lambda + 1^n)/\lambda$. It has been shown in [Haglund et al. 05b] that each of the terms $\sum_\tau q^{\text{dinv}(\tau)} \mathbf{z}_\tau$ is a symmetric function. See [Haglund 08] for more on all this.

Chapter 11

Coinvariant-Like Spaces

Many results of a flavor similar to those of Chapters 8 and 10 have been obtained in recent years. However, the picture they paint is often incomplete and much remains to be done. We give a sampling of these results here, as well as some related open problems and conjectures.

11.1 Operator-Closed Spaces

When we consider the characterization of the space $\mathcal{D}_n$ of diagonal harmonics as the smallest vector space $\mathcal{L}_{\partial,E}[\Delta_n(\mathbf{x})]$ containing the Vandermonde determinant that is closed under partial derivatives and applications of the operators E_k (see Section 10.5), one question naturally comes to mind. What is the structure of the space $\mathcal{L}_{\partial,E}[A]$, with A any spanning set for an invariant subspace of $\mathbb{C}[\mathbf{x}, \mathbf{y}]$? An example of this would be to choose A to be any one element set containing an antisymmetric polynomial such as $\Delta_\mu(\mathbf{x}, \mathbf{y})$. Observe that in this specific case, we would get an interesting subspace of $\mathcal{D}_n$.

Another possible direction is to start with an antisymmetric polynomial in $\mathbf{x}$. For instance we can choose this polynomial to be $e_k(\mathbf{x})\Delta_n(\mathbf{x})$. We denote by $\mathcal{D}_{n;k}$ the resulting bigraded $\mathfrak{S}_n$-module. Observe that for $k = 0$ this is simply the usual space of diagonal harmonics $\mathcal{D}_n$, so that Theorem 10.7 can trivially be reformulated as $\mathrm{Frob}_{qt}(\mathcal{D}_{n;0}) = \nabla(e_n)$. Less trivial is the fact that the space $\mathcal{D}_{n;n}$ can also be made entirely explicit. Building on the discussion of Section 10.3 and the observation that the Vandermonde-like determinant associated with the punctured diagram $(n + 1)/0, 0$ is none other than $e_n(\mathbf{x})\Delta_n(\mathbf{x})$, we can deduce that

$$\mathrm{Frob}_{qt}(\mathcal{D}_{n;n}) = s_1^\perp \nabla(e_{n+1}). \tag{11.1}$$

We can further calculate the right-hand side of this equation using operator identities that can be found in [Bergeron et al. 99a]. One considers the operators $\mathbf{D}_k$ on symmetric functions defined by the formal series identity

$$\sum_{k \in \mathbb{Z}} \mathbf{D}_k f(\mathbf{z}) y^k := f[\mathbf{z} + \theta/y] E[-y],$$

with $\theta = (1-t)(1-q)$. One of the identities derived in [Bergeron et al. 99a] states that $\nabla^{-1} s_1^{\perp} \nabla = \frac{1}{\theta} \mathbf{D}_{-1}$, hence we have

$$s_1^{\perp} \nabla(e_{n+1}) = \nabla\left(\frac{1}{\theta} \mathbf{D}_{-1} e_{n+1}\right). \tag{11.2}$$

We can now calculate, using (4.6), that

$$\frac{1}{\theta} \mathbf{D}_{-1} e_{n+1}(\mathbf{z}) = \frac{1}{\theta} e_{n+1}[\mathbf{z} + \theta/y] E[-y] \Big|_{y^{-1}}$$

$$= \sum_{j=0}^{n} e_j(\mathbf{z}) \frac{e_{n+1-j}[\theta]}{\theta y^{n+1-j}} E[-y] \Big|_{y^{-1}}$$

$$= \sum_{j=0}^{n} e_j(\mathbf{z}) e_{n-j}(\mathbf{z}) \frac{(-1)^{n-j} e_{n+1-j}[\theta]}{\theta}.$$

Using plethystic formulas, we can now evaluate $e_{n+1-j}[\theta]$ (see (3.28)) to get

$$\frac{(-1)^{n-j} e_{n+1-j}[\theta]}{\theta} = [n - j + 1]_{q,t},$$

with the notation $[m]_{q,t} = q^{m-1} + q^{m-2}t + \cdots qt^{m-2} + t^{m-1}$. In view of (11.1) and (11.2) we get

$$\mathrm{Frob}_{qt}(\mathcal{D}_{n;n}) = \sum_{j=0}^{n} [n - j + 1]_{q,t} \nabla(e_j e_{n-j}). \tag{11.3}$$

Our experiments and results suggest that a simple formula of this flavor holds for all spaces $\mathcal{D}_{n;k}$.

Conjecture 11.1 *For all k between 0 and n,*

$$\mathrm{Frob}_{qt}(\mathcal{D}_{n;k}) = \sum_{j=0}^{k} [k - j + 1]_{q,t} \nabla(e_j e_{n-j}). \tag{11.4}$$

As we have seen, there is a natural link between the study of the spaces $\mathcal{D}_{n;k}$ and the spaces associated with punctured diagrams of Section 10.3.

Another striking aspect of Conjecture 11.1 is that it suggests a potential candidate for the representation-theoretic description of the apparently Schur-positive symmetric function $\nabla(-s_{21^{n-1}})$, if we are to believe Conjecture 9.1. Indeed, observing that (11.4) takes the form

$$\mathrm{Frob}_{qt}(\mathcal{D}_{n;1}) = \nabla(e_j e_{n-j}) + (q+t)\nabla(e_n)$$

when $k = 1$, and using the fact that $s_{21^{n-1}} = e_1 e_{n-1} - e_n$, we calculate directly that

$$\nabla(-s_{21^{n-1}}) = \mathrm{Frob}_{qt}(\mathcal{D}_{n;1}) - (q+t+1)\,\mathrm{Frob}_{qt}(\mathcal{D}_{n;0}). \qquad (11.5)$$

The following exercise implies that we have

$$\mathcal{D}_n \oplus e_n(\mathbf{x})\mathcal{D}_n \oplus e_n(\mathbf{y})\mathcal{D}_n \subseteq \mathcal{D}_{n;1}. \qquad (11.6)$$

Exercise. Show that the polynomials $\Delta(\mathbf{x})$ and $e_1(\mathbf{y})\Delta(\mathbf{x})$ both lie in $\mathcal{D}_{n;1}$. Considering the linear operator (see Section 10.2)

$$\partial X := p_{1,0}(\partial\mathbf{x}, \partial\mathbf{y}),$$

check that $\mathcal{D}_n$ is the image of $\mathcal{D}_n + e_n(\mathbf{x})\mathcal{D}_n + e_n(\mathbf{y})\mathcal{D}_n$ under ∂X, and that $\mathcal{D}_n + e_n(\mathbf{y})\mathcal{D}_n$ is the kernel of

$$\partial X : \big(\mathcal{D}_n + e_n(\mathbf{x})\mathcal{D}_n + e_n(\mathbf{y})\mathcal{D}_n\big) \longrightarrow \mathcal{D}_n.$$

Using a similar conclusion for $\partial Y := p_{0,1}(\partial\mathbf{x}, \partial\mathbf{y})$, conclude that (11.6) holds.

Denoting by $\mathcal{M}_n$ the orthogonal complement in $\mathcal{D}_{n;1}$ of the subspace $\mathcal{D}_n \oplus e_n(\mathbf{x})\mathcal{D}_n \oplus e_n(\mathbf{y})\mathcal{D}_n$, under the hypothesis (11.4) we get

$$\mathrm{Frob}_{qt}(\mathcal{M}_n) = \nabla(-s_{21^{n-1}}),$$

in view of (11.5). Similar descriptions can be obtained for $\nabla(-s_{2^k 1^{n-2k}})$.

11.2 Quasisymmetric Polynomials Modulo Symmetric Polynomials

Here, we consider similar situations in the chain of rings $\mathbb{C}[\mathbf{x}]^{\sim_r \mathfrak{S}_n}$, of r-quasisymmetric polynomials[1] that sit in between the smaller ring of symmetric polynomials and the global ring of polynomials:

$$\mathbb{C}[\mathbf{x}]^{\mathfrak{S}_n} \hookrightarrow \cdots \hookrightarrow \mathbb{C}[\mathbf{x}]^{\sim_r \mathfrak{S}_n} \hookrightarrow \cdots \hookrightarrow \mathbb{C}[\mathbf{x}]^{\sim \mathfrak{S}_n} \hookrightarrow \mathbb{C}[\mathbf{x}]. \qquad (11.7)$$

---
[1] See Chapter 4.

All these inclusions are graded ring monomorphisms. In 2002, it was conjectured by C. Reutenauer and the author that $\mathbb{C}[\mathbf{x}]^{\sim \mathfrak{S}_n}$ is a free $\mathbb{C}[\mathbf{x}]^{\mathfrak{S}_n}$-module, and it was shown that this implies the Hilbert series of $\mathbb{C}[\mathbf{x}]^{\sim \mathfrak{S}_n}$ is given by the formula

$$\mathrm{Hilb}_q(\mathbb{C}[\mathbf{x}]^{\sim \mathfrak{S}_n}) = \frac{\varphi_n(q)}{\prod_{i=1}^n 1 - q^i}, \tag{11.8}$$

where the $\varphi_n(q)$ are the polynomials characterized by the recurrence

$$\varphi_n(q) = \varphi_{n-1}(q) + q^n\big([n]_q! - \varphi_{n-1}(q)\big), \quad \text{whenever } n \geq 1 \tag{11.9}$$

with initial condition $\varphi_0(q) = 1$. One easily verifies recursively that both $\varphi_n(q)$ and $[n+1]_q! - \varphi_n(q)$ have positive integer coefficients, since

$$[n+1]_q! - \varphi_n(q) = (1 + \cdots + q^n)[n]_q! - \big(\varphi_{n-1}(q) + q^n\big([n]_q! - \varphi_{n-1}(q)\big)\big)$$
$$= \big([n]_q! - \varphi_{n-1}(q)\big) + (q + \cdots + q^{n-1})[n]_q! + q^n\varphi_{n-1}(q).$$

The Bergeron–Reutenauer conjecture was rapidly settled in the paper [Garsia and Wallach 03]. Later, it was conjectured in [Hivert 00] that $\mathbb{C}[\mathbf{x}]^{\sim_r \mathfrak{S}_n}$ is also a free $\mathbb{C}[\mathbf{x}]^{\mathfrak{S}_n}$-module for all r. Hivert showed that this would imply

$$\mathrm{Hilb}_q(\mathbb{C}[\mathbf{x}]^{\sim_r \mathfrak{S}_n}) = \frac{\varphi_n^{(r)}(q)}{\prod_{i=1}^n 1 - q^i}, \tag{11.10}$$

with $\varphi_n^r(q)$ the positive integer coefficient polynomial given by the very explicit formula

$$\varphi_n^{(r)}(q) = \sum_{\sigma \in \mathfrak{S}_n} q^{\mathrm{maj}(\sigma) + r(n - \mathrm{fix}(\sigma))}. \tag{11.11}$$

History almost repeated itself when Hivert's conjecture was rapidly shown to be true by Wallach. We present here a finer conjecture formulated by N. Bergeron and the author in [Bergeron and Bergeron 05]. The starting point of this approach is another recursive description of the polynomials φ_n^r:

(1) $\varphi_n^{(r)}(q) = \varphi_{n-1}^{(r)}(q) + \psi_n^{(r)}(q)$

(2) $\psi_n^{(r)}(q) = (1 + q + \cdots + q^{n-2})\big(q^{2r+n-1}\varphi_{n-2}^{(r)}(q) + q^r\psi_{n-1}^{(r)}(q)\big),$

with $\varphi_0^{(r)}(q) = \varphi_1^{(r)}(q) = 1$, and $\psi_0^{(r)}(q) = \psi_1^{(r)}(q) = 0$. As we will see, it is natural to add a new formal parameter and instead consider the recurrence

(1) $\Phi_n(q,t) = \Phi_{n-1}(q,t) + \Psi_n(q,t)$

(2) $\Psi_n(q,t) = (q + q^2 + \cdots + q^{n-1})\big(t^2q^n\Phi_{n-2}(q,t) + t\Psi_{n-1}(q,t)\big),$

$$\tag{11.12}$$

with $\Phi_0(q,t) = \Phi_1(q,t) = 1$ and $\Psi_0(q,t) = \Psi_1(q,t) = 0$. It is not hard to show, in view of (11.10), that the recurrence affords the solution

$$\Phi_n(q,t) = \sum_{\sigma \in \mathfrak{S}_n} q^{\mathrm{maj}(\sigma)+n-\mathrm{fix}(\sigma)} t^{n-\mathrm{fix}(\sigma)}.$$

Clearly, we can compute $\varphi_n^{(r)}(q)$ from $\Phi_n(q,t)$ as

$$\varphi_n^{(r)}(q) = \Phi_n(q, q^{r-1}).$$

The first few values of $\Phi_n(q,t)$ are

$\Phi_0(q,t) = 1,$

$\Phi_1(q,t) = 1,$

$\Phi_2(q,t) = t^2 q^3 + 1,$

$\Phi_3(q,t) = (t^3 + t^2)q^5 + (t^3 + t^2)q^4 + t^2 q^3 + 1,$

$\Phi_4(q,t) = t^4 q^{10} + t^4 q^9 + (2t^4 + t^3)q^8 + (2t^4 + 2t^3 + t^2)q^7$
$\qquad + (2t^4 + 2t^3 + t^2)q^6 + (t^4 + 2t^3 + 2t^2)q^5$
$\qquad + (t^3 + t^2)q^4 + t^2 q^3 + 1.$

To further expand our discussion, let us introduce the ideal $\mathcal{J}_n$ of $\mathbb{C}[\mathbf{x}]^{\sim_r \mathfrak{S}_n}$ generated by the symmetric polynomials without constant term. Among the implications of Hivert's conjecture is the fact that the quotient ring

$$\mathcal{QC}_n^r := \mathbb{C}[\mathbf{x}]^{\sim_r \mathfrak{S}_n}/\mathcal{J}_n$$

has dimension $n!$. Moreover, $\mathcal{J}_n$ is homogeneous, the space $\mathcal{QC}_n^r$ is graded, and the conjecture implies that $\mathrm{Hilb}_q(\mathcal{QC}_n^r) = \varphi_n^{(r)}(q)$. Observe that Hivert's conjecture is equivalent to the existence of a (graded) isomorphism

$$\mathbb{C}[\mathbf{x}]^{\sim_r \mathfrak{S}_n} \simeq \mathbb{C}[\mathbf{x}]^{\mathfrak{S}_n} \otimes \mathcal{QC}_n^r. \tag{11.13}$$

Due to the parallelism between this statement and (8.1), it seems natural to call $\mathcal{QC}_n^r$ the *quasisymmetric coinvariant space*.

Partition-Free Reduction

Our aim is now to show that $\mathbb{C}[\mathbf{x}]^{\sim_r \mathfrak{S}_n}$ is actually spanned by elements of the form $M_\mathbf{b} m_\lambda$ with $\mathbf{b} > r$ and λ a partition with part sizes at most r.[2] Clearly $M_\mathbf{b} m_\lambda$ belongs to the ideal $\mathcal{J}_n$ except when λ is the zero partition. It thus follows that $\mathbb{C}[\mathbf{x}]^{\sim_r \mathfrak{S}_n}/\mathcal{J}_n$ is generated by the partition-free quasisymmetric polynomials.

---

[2]For notations and definitions, see Section 4.12.

Lemma 11.2. *Every r-quasisymmetric monomial $M_{\mathbf{a},\mu}$ can be written in the form $\sum_{\mathbf{b},\lambda} \gamma_{\mathbf{b},\lambda} M_{\mathbf{b}} m_\lambda$ for some scalars $\gamma_{\mathbf{b},\lambda} \in \mathbb{Q}$ with $\mathbf{b} > r$ and $0 \le \lambda_i \le r - 1$. Thus, the set of $M_{\mathbf{b}} m_\lambda$ with $|\mathbf{b}| + |\lambda| = d$ spans the degree-d homogeneous component of $\mathbb{C}[\mathbf{x}]^{\sim_r \mathfrak{S}_n}$.*

Proof: We prove this by recurrence on the length of μ. Let $\mathbf{a} = (a_1, \ldots, a_k)$, and observe that the case $\ell(\mu) = 0$ is trivially true. Let $\mu = (\mu_1 \cdots \mu_{n-1}, b)$ and set $\nu := (\mu_1 \cdots \mu_{n-1})$. Then, if we denote by $\nu +_j b$ the partition obtained (up to reordering) by adding b to the jth part of ν and likewise by $\mathbf{a} +_i b$ the composition obtained (without reordering) by adding b to the ith part of $\mathbf{a}$, the multiplication rule for r-quasisymmetric polynomials (see [Hivert 08]) gives

$$M_{\mathbf{a},\nu} M_b = M_{\mathbf{a},\mu} + \sum_{i=1}^{\ell(\mathbf{a})} M_{\mathbf{a}+_i b,\nu} + \sum_{j=1}^{n-1} M_{\mathbf{a},\nu+_j b}. \qquad (11.14)$$

Rewriting (11.14) to isolate $M_{\mathbf{a},\mu}$ on the left-hand side, we get

$$M_{\mathbf{a},\mu} = M_{\mathbf{a},\nu} M_b - \sum_{i=1}^{\ell(\mathbf{a})} M_{\mathbf{a}+_i b,\nu} - \sum_{j=1}^{n-1} M_{\mathbf{a},\nu+_j b}.$$

The inductive hypothesis can clearly be applied to each of the terms in the resulting right-hand side, so we are done with our proof. $\qquad \square$

Basis Conjecture

We now describe a conjecture for a uniform construction of a basis for the spaces $\mathcal{QC}_n^r$, with fixed n and $1 \le r$. For $m \in \mathbb{N}$, we consider compositions

$$\mathbf{a} + m := (a_1 + m, a_2 + m, \ldots, a_k + m)$$

obtained from a composition $\mathbf{a} = (a_1, a_2, \ldots, a_k)$. Let us now propose a rule for the construction of a basis of $\mathcal{QC}_n^r$ in the form of a set $\{M_{\mathbf{a}}\}_{\mathbf{a} \in \mathcal{A}_n^r}$, with $\mathcal{A}_n^r$ a well chosen set of compositions. Writing $\mathcal{A}_n = \mathcal{A}_n^1$, we conjecture that it is possible to choose $\mathcal{A}_n^r$ in such a way that $\mathcal{A}_n^r = \{\mathbf{a} + (r-1) \mid \mathbf{a} \in \mathcal{A}_n\}$ while respecting the equality $\sum_{\mathbf{a} \in \mathcal{A}_n^r} q^{|\mathbf{a}|} = \varphi_n^r(q)$. Indeed, to ensure that we get the right Hilbert series for all $\mathcal{A}_n^r$ constructed from $\mathcal{A}_n$ as above, it is sufficient that we have

$$\sum_{\mathbf{a} \in \mathcal{A}_n} q^{|\mathbf{a}|} t^{\ell(\mathbf{a})} = \Phi_n(q, t). \qquad (11.15)$$

When (11.15) holds we say that the set $\mathcal{A}_n$ is *length adequate*. Let us recursively define a length-adequate set $\mathcal{A}_n$ of compositions making use of auxiliary sets $\mathcal{B}_n$, by

(1) $\quad \mathcal{A}_n := \mathcal{A}_{n-1} + \mathcal{B}_n,$

(2) $\quad \mathcal{B}_n := \{n \cdot k \cdot \mathbf{a} \mid \mathbf{a} \in \mathcal{A}_{n-2}, 1 \leq k \leq m - 1\} \hspace{2cm} (11.16)$

$\quad\quad\quad + \{k \cdot \mathbf{b} \mid \mathbf{b} \in \mathcal{B}_{n-1}, 1 \leq k \leq m - 1\},$

with $\mathcal{A}_0 = \mathcal{A}_1 := \{0\}$, $\mathcal{B}_0 = \mathcal{B}_1 := \emptyset$, and addition between sets corresponding to disjoint union. Clearly, recurrence (11.16) exactly follows the form of recurrence (11.12), with t as a marker for the number of parts, and q for the size of the partition, hence $\mathcal{A}_n$ is indeed length adequate. Using the conventions that $a = r$, $b = r + 1$, $c = r + 2$, etc., that compositions are written as words, and that a set of compositions is written as a sum of words, we have for small values of n that

- $A_2 = 0 + ba,$

- $A_3 = A_2 + (ca + cb) + (aba + bba),$

- $A_4 = A_3 + (da + db + dc + daba + dbba + dcba) + (aca + bca + cca + acb + bcb + ccb + aaba + baba + caba + abba + bbba + cbba).$

Computer experiments for small values of n suggest that the set $\{M_\mathbf{a}\}_{\mathbf{a} \in \mathcal{A}_n^r}$ forms a linear basis of the space $\mathcal{QC}_n^r$. We have already mentioned that Wallach has shown that Hivert's conjecture is true, but his techniques do not produce a basis like the one described here. The problem of showing that $\{M_\mathbf{a}\}_{\mathbf{a} \in \mathcal{A}_n^r}$ is a linearly independent set is still open. This would clearly suffice to show that it is a basis.

11.3 Super-Coinvariant Space

By analogy with the $\mathfrak{S}_n$-coinvariant space, we consider here the graded *super-coinvariant space* $\mathcal{SC}_n := \mathbb{C}[\mathbf{x}]/\mathcal{J}_n$, where $\mathcal{J}_n$ is the ideal generated by constant-term-free quasisymmetric polynomials in the variables $\mathbf{x}$. Naturally, we also consider the isomorphic graded space $\mathcal{SH}_n$ of *super-harmonic polynomials* obtained as the orthogonal complement of $\mathcal{J}_n$. Observe that $\mathcal{SH}_n$ is a subspace of the space $\mathcal{H}_n$ of harmonic polynomials, since the ring of symmetric polynomials is a subring of the ring of quasisymmetric polynomials. We will now construct a basis of $\mathcal{SC}_n$ by giving an explicit set of monomial representatives naturally indexed by Dyck paths of length n. We obtain the following theorem, whose detailed proof can be found in [Aval et al. 04].

Theorem 11.3. *The dimension of $\mathcal{SH}_n$ (or $\mathcal{SC}_n$) is given by the nth Catalan number,*

$$\dim \mathcal{SH}_n = \dim \mathcal{SC}_n = \frac{1}{n+1}\binom{2n}{n}. \hspace{2cm} (11.17)$$

In fact, taking into account the grading by degree, our basis will make it clear that we have the Hilbert series formula

$$\mathrm{Hilb}_t(\mathcal{SH}_n) = \sum_{k=0}^{n-1} \frac{n-k}{n+k}\binom{n+k}{k}t^k. \tag{11.18}$$

The following easily derived property of the fundamental basis of quasi-symmetric polynomials plays a crucial role in the construction that we intend to describe. For a composition $\mathbf{a} = (a_1, a_2, \ldots, a_k)$, setting $\mathbf{b} = (a_2, \ldots, a_k)$ and taking $\mathbf{y}$ such that $\mathbf{x} = x_1 + \mathbf{y}$, the following identity holds:

$$Q_{\mathbf{a}}(\mathbf{x}) = \begin{cases} x_1 Q_{\mathbf{a}-\mathbf{e}_1}(\mathbf{x}) + Q_{\mathbf{a}}(\mathbf{y}) & \text{if } a_1 > 1, \\ x_1 Q_{\mathbf{b}}(\mathbf{y}) + Q_{\mathbf{a}}(\mathbf{y}) & \text{otherwise.} \end{cases} \tag{11.19}$$

Here $\mathbf{e}_1$ stands for the length k vector $(1, 0, \ldots, 0)$, so that $\mathbf{a} - \mathbf{e}_1 = (a_1 - 1, a_2, \ldots, a_k)$.

A Basis for the Space of Polynomials

To any vector $\mathbf{v}$ in $\mathbb{N}^n$ let us associate a lattice path $\gamma(\mathbf{v})$ in $\mathbb{N} \times \mathbb{N}$ just as in Section 4.5 so that the monomial corresponding to the weight of the path is precisely $\mathbf{x}^{\mathbf{v}}$. We distinguish two kinds of paths (or of monomials) depending on whether or not they cross the diagonal $y = x$. If the path always remains above the diagonal, we say that it is a *Dyck path*[3] and that the corresponding monomial is *Dyck*. When a path (or the corresponding monomial) is not Dyck we say that it is *transdiagonal*. For example, the monomial $x_3 x_4^2 x_6$ is Dyck, whereas $x_2^2 x_3 x_5^2 x_6^2$ is transdiagonal. The monomial $\mathbf{x}^{\mathbf{v}}$ is transdiagonal for $v = (v_1, \ldots, v_n)$ if and only if there is some $1 \le m \le n$ such that $m < v_1 + \cdots + v_m$.

Now, for a vector $\mathbf{v}$ in $\mathbb{N}^n$, let us consider the leftmost nonzero coordinate a (if any) that follows a zero coordinate of $\mathbf{v}$, so that $\mathbf{v}$ has the form $(v_1, \ldots, v_k, 0, \ldots, 0, a, w_1, \ldots, w_\ell)$ with the v_i nonzero. Let us write

$$\mathbf{v}' = (v_1, \ldots, v_k, 0, \ldots, a, w_1, \ldots, w_\ell, 0),$$
$$\mathbf{v}'' = (v_1, \ldots, v_k, 0, \ldots, a - 1, w_1, \ldots, w_\ell, 0).$$

We recursively define a $\mathbf{v}$-indexed element $G_{\mathbf{v}}(\mathbf{x})$ of $\mathbb{C}[\mathbf{x}]$ as follows. If no coordinate of $\mathbf{v}$ follows a zero then we identify $\mathbf{v}$ with the composition obtained by erasing the zero in $\mathbf{v}$ and set $G_{\mathbf{v}}(\mathbf{x}) = Q_{\mathbf{v}}(\mathbf{x})$. Otherwise, we recursively set

$$G_{\mathbf{v}}(\mathbf{x}) := G_{\mathbf{v}'}(\mathbf{x}) + x_k G_{\mathbf{v}''}(\mathbf{x}),$$

---

[3]This is slightly different from our definition in Chapter 1.

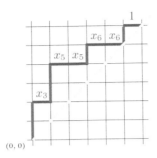

Figure 11.1. A Dyck monomial.

with k such that a sits in position $k + 1$ in $\mathbf{v}$. For instance, we have

$$G_{1020}(\mathbf{x}) = G_{1200}(\mathbf{x}) - x_2 G_{1100}(\mathbf{x})$$
$$= Q_{12}(\mathbf{x}) - x_2 Q_{11}(\mathbf{x}).$$

The key feature of these $G_{\mathbf{v}}$ is that the leading monomial of $G_{\mathbf{v}}$ (in lexicographic order) is $\mathbf{x}^{\mathbf{v}}$.[4] Hence $G_{\mathbf{v}}$ is a linear basis for the ring of polynomials $\mathbb{C}[\mathbf{x}]$. The second fact, which is a little harder to prove, is that the ideal $\mathcal{J}_n$ is precisely the linear span of the polynomials $G_{\mathbf{v}}$ for which the lattice path associated with $\mathbf{v}$ is transdiagonal. It follows that the super-coinvariant space $\mathcal{SC}_n$ has as basis the set of Dyck monomials in the variables $\mathbf{x}$. The lattice path corresponding to a Dyck monomial can always be completed in a more classical Dyck path by adding (as in Figure 11.1) some final horizontal steps at height n, giving them weight 1 each.

Exercise. Show that the number of Dyck monomials of degree k is given by the expression $\frac{n-k}{n+k}\binom{n+k}{k}$.

Diagonal Version

With the intent of extending these considerations to a diagonal setup, let us say that $C = (\mathbf{a}, \mathbf{b})$ is a length-k *bicomposition* of (m, ℓ) if $\mathbf{a} = (a_1, \ldots, a_k)$ and $\mathbf{b} = (b_1, \ldots, b_k)$ are both in $\mathbb{N}^k$, with $a_1 + \cdots + a_k = m$, $b_1 + \cdots + b_k = \ell$, and $a_i + b_i > 0$ for all $1 \le i \le k$.[5] Each (a_i, b_i) is said to be a *part* of the bicomposition C, and the integer $a_i + b_i$ is its *part size*. We extend to this context the vector exponential notation, writing X^C for the monomial $\mathbf{x^a y^b}$. Here $X = \{(x_1, y_1), \ldots, (x_n, y_n)\}$ is to be considered as an ordered set of variable pairs, with $\mathbf{x} = x_1, \ldots, x_n$ and $\mathbf{y} = y_1, \ldots, y_n$. With these

---

[4]For more on this see [Aval et al. 04].
[5]Notice that we allow $a_i = 0$ and $b_i = 0$, so that $\mathbf{a}$ and $\mathbf{b}$ are not compositions.

conventions at hand, let C be a length-k bicomposition and define the *monomial diagonally quasisymmetric polynomial* $\mathbb{M}_C(X)$ to be

$$\mathbb{M}_C(X) := \sum_{Y \subseteq X} Y^C, \qquad (11.20)$$

where the sum is over the k-subsets Y of X with order induced from that on X. Here is an example of this definition:

$$\mathbb{M}_{(2,0)(1,1)(0,2)}(X) = x_1^2 x_2 y_2 y_3^2 + x_1^2 x_2 y_2 y_4^2 + \cdots + x_{n-2}^2 x_{n-1} y_{n-1} y_n^2.$$

The linear span of all monomial diagonally quasisymmetric polynomials is closed under multiplication (see [Aval et al. 07]), and we denote by $\mathbb{C}[\mathbf{x}, \mathbf{y}]^{\sim \mathfrak{S}_n}$ the resulting ring of *diagonally quasisymmetric polynomials*. Consider now $\mathcal{DJ}_n$, the bihomogeneous ideal generated by diagonally quasisymmetric polynomials without constant term, and define the *diagonal super-coinvariant space*

$$\mathcal{DSC}_n := \mathbb{C}[\mathbf{x}, \mathbf{y}]/\mathcal{DJ}_n.$$

Naturally, we also consider the space of *diagonal super-harmonic polynomials*, defined as the orthogonal complement $\mathcal{DSH}_n := \mathcal{DJ}_n^\perp$. As in previous situations, these are two isomorphic bigraded spaces. The following conjecture for the bigraded Hilbert series of the space $\mathcal{DSC}_n$ is discussed further in [Aval et al. 07].

Conjecture 11.4 *Let* $\mathrm{Hilb}_{n+1}(i,j)$ *denote the coefficient of* $q^i t^j$ *in the bigraded Hilbert series (polynomial) of the space* $\mathcal{DSC}_{n+1}$. *Then* $\mathrm{Hilb}_{n+1}(i,j)$ *is the nth Catalan number when* $i = j$:

$$\mathrm{Hilb}_{n+1}(i,i) = \frac{1}{n+1}\binom{2n}{n}.$$

All the other nonzero coefficients can be recursively obtained as

$$\mathrm{Hilb}_{n+1}(i,j) = \sum_{i' \geq i, \, j' \leq j} \mathrm{Hilb}_n(i',j'), \qquad (11.21)$$

with $i > j$.

To emphasize the nice symmetries of the conjectured Hilbert polynomials, we resort to the matrix notation of Section 10.5 (see the example immediately following (10.19)). For instance, for $1 \leq n \leq 5$, the Hilbert matrices Hilb_n described in the conjecture are

$$\mathrm{Hilb}_1 = \begin{pmatrix} 1 \end{pmatrix}, \quad \mathrm{Hilb}_2 = \begin{pmatrix} 1 & \\ 1 & 1 \end{pmatrix}, \quad \mathrm{Hilb}_3 = \begin{pmatrix} 2 & & \\ 2 & 2 & \\ 1 & 2 & 2 \end{pmatrix},$$

$$\mathrm{Hilb}_4 = \begin{pmatrix} 5 & & & \\ 5 & 5 & & \\ 3 & 7 & 5 & \\ 1 & 3 & 5 & 5 \end{pmatrix}, \quad \mathrm{Hilb}_5 = \begin{pmatrix} 14 & & & & \\ 14 & 14 & & & \\ 9 & 24 & 14 & & \\ 4 & 14 & 24 & 14 & \\ 1 & 4 & 9 & 14 & 14 \end{pmatrix}.$$

The values in the first column (as well as those of the last row) of Hilb_n are clearly the coefficients of $\mathrm{Hilb}_t(\mathcal{SH}_n)$ given in (11.18). We can in fact construct a set $\mathcal{B}_n$ of monomials that we conjecture to be a linear basis of $\mathcal{DSC}_n$. As should be expected, this set contains two copies of the basis for $\mathcal{SH}_n$; one in the **x**-variables and one in the **y**-variables. The degrees of monomials in $\mathcal{B}_n$ are all less than or equal to $n - 1$. Moreover, the construction will make clear that the bigraded enumeration of $\mathcal{B}_n$ agrees with 11.4. For $i + j < n - 1$ or $(i, j) = (n - 1, 0)$ we recursively construct a set $\mathcal{B}_n(i, j)$ of monomials of bidegree (i, j),

$$\mathcal{B}_n(i, j) := \sum_{a \leq i, \, b \leq j} \mathcal{B}_{n-1}(a, b) \cdot x_n^{i-a} y_n^{j-b}, \tag{11.22}$$

with the sum standing for disjoint union, and a product between a set B and a monomial m representing the set of all products bm, $b \in B$. To generate the remaining components of $\mathcal{B}_n$, corresponding to $i + j = n - 1$ and $j \geq 1$, we introduce an operation ϕ on bicompositions defined by $\phi\big((\mathbf{a}, \mathbf{c})\big) = (\mathbf{a} - \mathbf{e}_m, \mathbf{b} + \mathbf{e}_m)$, where m is the smallest index such that $a_m > 0$ and $\mathbf{e}_m$ is the mth unit vector. Then $\Phi(X^C) = X^{\phi(C)}$ and

$$\mathcal{B}_n(n - i - 1, i) := \{\Phi(X^C) \mid X^C \in \mathcal{B}_n(n - i, i - 1)\}. \tag{11.23}$$

We then conjecture (see [Aval et al. 07]) that the set $\mathcal{B}_n$ is a linear basis of $\mathcal{DSC}_n$. For $1 \leq n \leq 3$, we get (in matrix-like format)

$$\mathcal{B}_1 = \{1\},$$

$$\mathcal{B}_2 = \left\{ \begin{matrix} y_2 & \\ 1 & x_2 \end{matrix} \right\},$$

$$\mathcal{B}_3 = \left\{ \begin{matrix} y_3^2, y_2 y_3 & & \\ y_3, y_2 & y_3 x_3, y_2 x_3 & \\ 1 & x_3, x_2 & x_3^2, x_2 x_3 \end{matrix} \right\}.$$

Exercise. Formulate a similar conjecture for *diagonally r-quasisymmetric polynomials*, which can be straightforwardly defined using the notions outlined in this section.

11.4   More Open Questions

There are many interesting variations on the central theme of this chapter that have been consider. With no pretension of being thorough, let us mention a few of these that seem to beg for more study or extension to diagonal contexts and/or other Coxeter/reflection groups.

m-Quasi-Invariant

An interesting analog of (8.1) considered in [Etingof and Strickland 03] and [Felder and Veselov 03] involves the ring $\mathcal{QI}_m[\mathbf{x}]$ of m-*quasi-invariants*, defined for some fixed integer m as the set of polynomials $f(\mathbf{x})$ such that $f(\mathbf{x}) - s_{ij}f(\mathbf{x})$ is divisible by $(x_i - x_j)^{2m+1}$ for all $1 \leq i < j \leq n$. Recall here that the effect of the permutation s_{ij} on $f(\mathbf{x})$ is simply to transpose the variables x_i and x_j. Since we have $f(\mathbf{x}) - s_{ij}f(\mathbf{x}) = 0$ when f is symmetric, all symmetric polynomials are m-quasi-invariant. It is shown in [Felder and Veselov 03] that the quotient $\mathcal{QR}_n := \mathcal{QI}_m[\mathbf{x}]/(e_1, \ldots, e_n)$, is a graded version of the left-regular representation of $\mathfrak{S}_n$ whose Frobenius characteristic is

$$\mathrm{Frob}_t(\mathcal{QR}_n) = \sum_{\mu \vdash n}\left(\sum_{\lambda(\tau)=\mu} t^{\mathrm{ch}(\tau)+m(\binom{n}{2}-n(\mu)+n(\mu'))}\right)s_\mu, \qquad (11.24)$$

where τ varies in the set of standard tableaux of shape μ. It is also shown in [Etingof and Strickland 03] that $\mathcal{QI}_n$ is a $\mathbb{C}[\mathbf{x}]^{\mathfrak{S}_n}$-module.

Deformed Steenrod Operators

Although still conjectural (see [Hivert and Thiéry 04]), yet another interesting variation relates to the $\mathfrak{S}_n$-module $\mathcal{H}_{n;q}$ of common polynomial zeros of the operators

$$D_{k;q} := \sum_{i=1}^{n} qx_i\partial x_i^{k+1} + \partial x_i^k.$$

Evidently the resulting space coincides with the space of $\mathfrak{S}_n$-harmonic polynomials when $q = 0$. But the striking fact is that $\mathcal{H}_{n;q}$ appears always to be isomorphic to $\mathcal{H}_n$ as a graded $\mathfrak{S}_n$-module whenever $q \geq 0$. It is also noteworthy that $\mathcal{H}_{n;q}$ is simply the intersection of the kernels of $D_{1;q}$ and $D_{2;q}$, since we have

$$D_{k;q}D_{j;q} - D_{j;q}D_{k;q} = q(k - j)D_{k+j;q}$$

so that any polynomial in the kernel of both $D_{1;q}$ and $D_{2;q}$ must necessarily be in the kernel of $D_{k;q}$ for all $k \geq 1$. Among the open problems related to similar looking operators, we find a conjecture (see [Wood 98, Conjecture 7.3]) that is made more precise below. Our[6] story here will involve two families of operators, $\widetilde{D}_k := \sum_{i=1}^n x_i \partial_{x_i}^{k+1}$ and $\widehat{D}_k := \sum_{i=1}^n \partial_{x_i}^{k+1} x_i$, related by the nice operator identity $\widetilde{D}_k e_n(\mathbf{x}) = e_n(\mathbf{x}) \widehat{D}_k$. We emphasize that all of the terms are to be considered as operators; in particular, $e_n(\mathbf{x})$ is multiplication by $x_1 x_2 \cdots x_n$. To continue with our description, it will be useful to write $e_{\mathbf{y}}$ for the product of the variables in $\mathbf{y}$. We denote by $\widetilde{\mathcal{H}}_n = \widetilde{\mathcal{H}}_{\mathbf{x}}$ the space of common zeros of the $\widetilde{D}_k$, and by $\widehat{\mathcal{H}}_{\mathbf{x}}$ the space of common zeros of the $\widehat{D}_k$, both over the $\mathbf{x}$ of n variables. These are respectivly the spaces of *tilde harmonics* and *hat harmonics*. Direct computations show that

$\mathrm{Hilb}_t(\widetilde{\mathcal{H}}_2) = 1 + 2t + t^2 + t^3,$

$\mathrm{Hilb}_t(\widetilde{\mathcal{H}}_3) = 1 + 3t + 3t^2 + 4t^3 + 2t^4 + 2t^5 + t^6,$

$\mathrm{Hilb}_t(\widetilde{\mathcal{H}}_4) = 1 + 4t + 6t^2 + 10t^3 + 9t^4 + 11t^5 + 9t^6 + 6t^7 + 5t^8 + 3t^9 + t^{10}.$

We are now ready to state that the space of tilde harmonics has the direct sum decomposition

$$\widetilde{\mathcal{H}}_n = \bigoplus_{\mathbf{y} \subseteq \mathbf{x}} e_{\mathbf{y}} \widehat{\mathcal{H}}_{\mathbf{y}}, \qquad (11.25)$$

where we take $\widehat{\mathcal{H}}_{\mathbf{y}} = \mathbb{C}$ when $\mathbf{y} = \emptyset$. Thus, we get

$$\mathrm{Frob}_t(\widetilde{\mathcal{H}}_n) = 1 + \sum_{k=1}^n t^k F_k(\mathbf{z}; t) h_{n-k}(\mathbf{z}), \qquad (11.26)$$

with $F_k(\mathbf{z}; t) = \mathrm{Frob}_t(\widehat{\mathcal{H}}_{\mathbf{y}})$ for any set $\mathbf{y}$ of k variables. This makes sense since the Frobenius characteristic in question is evidently independent of the actual choice of variables. Equation (11.25) is easily derived from the operator identity relating the two sets of operators. If the following conjecture holds, (11.25) also implies a more explicit version of Wood's conjecture:

Conjecture 11.5 (Bergeron and Garsia) *The ordinary space of $\mathfrak{S}_n$-harmonics is isomorphic to $\widehat{\mathcal{H}}_{\mathbf{x}}$ as a graded $\mathfrak{S}_n$-module.*

Diagonal Version

We have been led to suspect that there is similar behavior for deformation of operators in the context of diagonal harmonics. Indeed, our experiments

---

[6]This is joint work with A. Garsia.

suggest that the bigraded Hilbert series of the space corresponding to the set of common zeros of the operators $\sum_{i=1}^{n} a_i \partial_{x_i}^k \partial_{y_i}^j$, for $k, j \in \mathbb{N}$, $k + j > 0$, coincides with the bigraded Hilbert series of the space of diagonal harmonic polynomials for $\mathfrak{S}_n$ whenever $\sum_{k \in K} a_k \neq 0$ for all nonempty subsets K of $\{1, \ldots, n\}$. Interesting aspects of these questions arise when we choose the parameters a_i to have some symmetry. Thus, if the vector $\mathbf{a} = (a_1, \ldots, a_n)$ is fixed by some group G, we get a G-module structure on the spaces considered, for which we now need to find the decomposition into irreducibles.

More Sets of Variables

All the spaces considered in Chapters 10 and 11 involve either one or two sets of variables. Recent work of the author suggests that a beautiful global theory lies behind a generalization to an arbitrary number of sets of variables. The relevant multi-graded Hilbert series and Frobenius characteristics have coefficients that are symmetric in the parameters used to mark the various sets of variables, i.e., those that replace q and t. These symmetric functions seem to have uniform descriptions, which is to say that they may be expressed in a format that is independent of the number of parameters. In particular, it follows that we get nice new decompositions of the bigraded Hilbert series of diagonal harmonics. Even if we leave the explanation of these recent developments to upcoming papers, let us give the relevant decompositions for small values of n:

$$\mathrm{Hilb}_{qt}(\mathcal{D}_1) = 1,$$
$$\mathrm{Hilb}_{qt}(\mathcal{D}_2) = 1 + h_1,$$
$$\mathrm{Hilb}_{qt}(\mathcal{D}_3) = 1 + 2\,h_1 + (h_1^2 + h_2) + h_3,$$
$$\mathrm{Hilb}_{qt}(\mathcal{D}_4) = 1 + 3\,h_1 + (3\,h_1^2 + 2\,h_2) + (h_1^3 + 3\,h_1 h_2 + 2\,h_3)$$
$$+ (4\,h_1 h_3 + h_4) + (h_1 h_4 + 2\,h_5) + h_6,$$

where each complete homogeneous symmetric function h_k is to be evaluated in the two variables q and t. The striking conjecture here is that these formulas are always h-positive and universal. This is to say that they hold for all (correctly defined) spaces of diagonal harmonics in k sets of variables, in which case we evaluate them at $q_1, \ldots, q_k$ to get the k-graded Hilbert series of the relevant space. In particular, this conjecture would imply that we should have a general formula for $\mathcal{H}_n$ of the form

$$\mathcal{H}_n = \sum_{\sigma \in \mathfrak{S}_n} h_{\mu(\sigma)},$$

with $\mu(\sigma)$ some partition of the inversion number $\ell(\sigma)$ of σ. This would assure that they specialize to the q-analog of $n!$ in the case $k = 1$.

Appendix A

Formulary

In the course of working with the Macdonald functions and related operators, we have found it useful to have at hand a list of the main formulas used. To help keep the readers on the same track, we have listed the formulas that appear in our discussions (as well as some others). This appendix may also serve as reference for our notations.

A.1 Some q-Identities

For a permutation σ, $\varepsilon(\sigma)$ is the *sign* of the permutation σ and it is equal to $(-1)^{\ell(\sigma)}$ with $\ell(\sigma)$ the *number of inversions* of σ. $\mathrm{Des}(\sigma)$ is the *set of descents* of σ, and $\mathrm{maj}(\sigma) := \sum_{i \in \mathrm{Des}(\sigma)} i$ is the *major index*. The partition giving the *cycle structure* of σ is $\lambda(\sigma)$ and $\gamma(\sigma)$ is the *number of cycles*. The *q-shifted factorial* $(a; q)_m$ is the product $(1-a)(1-aq) \cdots (1-aq^{m-1})$. The following q-identities (see Section 1.1) play a significant role in our story:

$$\sum_{\sigma \in \mathfrak{S}_n} q^{\ell(\sigma)} = \sum_{\sigma \in \mathfrak{S}_n} q^{\mathrm{maj}(\sigma)} = \prod_{k=1}^{n} \frac{1-q^k}{1-q}$$
$$= (1+q)(1+q+q^2) \cdots (1 + \cdots + q^{n-1}).$$

The right-hand side of this equation is often denoted by

$$[n]_q! := [1]_q [2]_q \cdots [n]_q,$$

where

$$[k]_q := \frac{1-q^k}{1-q} = 1 + q + \cdots + q^{k-1}$$

for $k > 0$, and $[0]_q := 1$.

The *q-binomial coefficients* are defined in (1.12) as

$$\begin{bmatrix} n + k \\ k \end{bmatrix}_q := \frac{[n]_q!}{[k]_q![n-k]_q!}.$$

This is a degree $k(n-k)$ polynomial in $\mathbb{N}[q]$, and we have (see (1.11))

$$\begin{bmatrix} n \\ k \end{bmatrix}_q = \sum_{\mu \subseteq n^k} q^{|\mu|},$$

as well as the *q-binomial identity*

$$\prod_{k=0}^{n-1}(1 + q^k z) = \sum_{k=0}^{n} q^{k(k-1)/2} \begin{bmatrix} n \\ k \end{bmatrix}_q z^k.$$

The q-binomial coefficients satisfy the recurrence (1.13)

$$\begin{bmatrix} m \\ k \end{bmatrix}_q = q^k \begin{bmatrix} m-1 \\ k \end{bmatrix}_q + \begin{bmatrix} m-1 \\ k-1 \end{bmatrix}_q,$$

with initial condition $\begin{bmatrix} m \\ k \end{bmatrix}_q = 1$ if $k = 0$ or $k = m$. Two q-analogs of the Catalan numbers play a role here. They are respectively characterized by (see (1.16) and (1.18))

$$\mathbf{C}_n(q) = \frac{1}{[n+1]_q} \begin{bmatrix} 2n \\ n \end{bmatrix}_q$$

and

$$\mathcal{C}_k(q) = \sum_{j=1}^{k} q^{j(k-j)} \mathcal{C}_{j-1}(q)\mathcal{C}_{k-j}(q),$$

with $\mathcal{C}_0(q) = 1$.

For many other formulas regarding q-analogs, see [Kac and Cheung 02].

A.2   Partitions and Tableaux

(See Sections 1.3 and 1.5.) For a partition μ of n and a cell $c = (i,j)$ in μ, the *arm length* of c is $a_\mu(c) := \mu_i - i$, and $\ell_\mu(c) := \mu'_j - j$ is its *leg length*. Further parameters are

$$n(\mu) = \sum_{i=1}^{n}(i-1)\mu_i = \sum_{(i,j)\in\mu} j = \sum_{c\in\mu} \ell_\mu(c),$$

$$n(\mu') = \sum_{i=1}^{n}(i-1)\mu'_i = \sum_{(i,j)\in\mu} i = \sum_{c\in\mu} a_\mu(s),$$

with μ' standing for the *conjugate* of μ. The *hook length formula* (2.1), giving the number of standard tableaux of shape μ, is

$$f^\mu = \frac{n!}{\prod_{c \in \mu} h(c)},$$

with the hook length $h(c)$ of a cell $c = (i, j)$ of μ equal to $(\mu_{j+1} - i - 1) + (\mu'_{i+1} - j - 1) + 1$. It affords as q-analog the formula (2.8):

$$\sum_{\lambda(\tau) = \mu} q^{\mathrm{coch}(\tau)} = q^{n(\mu)} \frac{(q; q)_n}{\prod_{c \in \mu}(1 - q^{h(c)})},$$

where $\mathrm{coch}(\tau)$ is the cocharge of the standard tableau τ. The Kostka number $K_{\lambda, \mu}$ (see Section 2.1) is the number of semi-standard tableaux of shape λ and content $\mu = 1^{\mu_1} 2^{\mu_2} \cdots k^{\mu_k}$. Its q-analog (see Section 2.6) is

$$K_{\lambda, \mu}(q) = \sum_\tau q^{\mathrm{ch}(\tau)},$$

summation being over standard tableaux τ of shape μ, with $\mathrm{ch}(\tau)$ the charge of τ. Charge and cocharge are linked by the relation $n(\lambda) = \mathrm{ch}(\tau) + \mathrm{coch}(\tau)$, for any tableau τ of shape μ.

Various q, t-parameters linked to partitions (see Section 1.3) are as follows:

$$T_\mu := t^{n(\mu)} q^{n(\mu')}, \quad B_\mu := \sum_{(i,j) \in \mu} q^i t^j, \quad \lambda_\mu := 1 - (1 - t)(1 - q) B_\mu,$$

as well as

$$\varepsilon_\mu = \prod_{c \in \mu}(q^{a_\mu(c)} - t^{\ell_\mu(c)+1}) \quad \text{and} \quad \varepsilon'_\mu = \prod_{c \in \mu}(t^{\ell_\mu(c)} - q^{a_\mu(c)+1}).$$

For ν covered by μ in Young lattice, the coefficients appearing in the following Pieri formula for integral form Macdonald functions H_μ (see Section A.4)

$$s_1^\perp H_\mu = \sum_{\nu \to \mu} c_{\mu\nu} H_\nu, \quad s_1 H_\mu = \sum_{\nu \to \mu} d_{\mu\nu} H_\mu,$$

are given by the expressions (see [Macdonald 95])

$$c_{\mu\nu} = \prod_{c \in \mathcal{R}_{\mu\nu}} \frac{t^{\ell_\mu(c)} - q^{a_\mu(c)+1}}{t^{\ell_\nu(c)} - q^{a_\nu(c)+1}} \prod_{c \in \mathcal{C}_{\mu\nu}} \frac{q^{a_\mu(c)} - t^{\ell_\mu(c)+1}}{q^{a_\nu(c)} - t^{\ell_\nu(c)+1}},$$

$$d_{\mu\nu} = \prod_{s \in \mathcal{R}_{\mu\nu}} \frac{q^{a_\nu(c)} - t^{\ell_\nu(c)+1}}{q^{a_\mu(c)} - t^{\ell_\mu(c)+1}} \prod_{s \in \mathcal{C}_{\mu\nu}} \frac{t^{\ell_\nu(c)} - q^{a_\nu(c)+1}}{t^{\ell_\mu(c)} - q^{a_\mu(c)+1}},$$

where $\mathcal{R}_{\mu\nu}$ (resp. $\mathcal{C}_{\mu\nu}$) stands for the row (resp. column) containing the corner μ/ν. Some of the identities relating these various parameters are

$$\varepsilon_\mu = (-1)^n T_\mu t^n \varepsilon_\mu(1/q, 1/t), \qquad \varepsilon_\mu' = (-1)^n T_\mu q^n \varepsilon_\mu'(1/q, 1/t),$$

$$T_\mu = e_n[B_\mu], \qquad\qquad \lambda_\mu = t^n + (1-t)\sum_{i=1}^{n} t^{i-1} q^{\mu_i}.$$

We also have $(1-t)(1-q)\varepsilon_\nu \varepsilon_\nu' c_{\mu\nu} = d_{\mu\nu}\varepsilon_\mu \varepsilon_\mu'$ and the less obvious identities (see [Bergeron and Garsia 99])

$$\sum_{\nu \to \mu} c_{\mu\nu}\left(\frac{T_\mu}{T_\nu}\right)^k = \begin{cases} (1-t)^{-1}(1-q)^{-1}h_{k+1}[-\lambda_\mu/qt] & \text{if } k \geq 1, \\ B_\mu & \text{if } k = 0, \end{cases}$$

$$\sum_{\mu \leftarrow \nu} d_{\mu\nu}\left(\frac{T_\mu}{T_\nu}\right)^k = \begin{cases} (-1)^{k-1}e_{k-1}[-\lambda_\nu] & \text{if } k \geq 1, \\ 1 & \text{if } k = 0, \end{cases}$$

where e_k and h_k stand for the elementary and complete homogeneous symmetric functions, and the brackets denote plethystic substitution (see Section A.3).

A.3   Symmetric and Antisymmetric Functions

(See Chapter 3 and [Macdonald 95].) The *complete homogeneous symmetric functions* are $h_d = \sum_{\lambda \vdash d} m_\lambda$, with m_λ denoting the *monomial symmetric functions*. They afford the generating function

$$H(t) := \sum_{d \geq 0} h_d t^k = \prod_{i=1}^{n} \frac{1}{1 - x_i t},$$

whereas for the *elementary symmetric functions*, $e_d = m_{11\cdots 1}$ (d copies of 1), we have

$$E(t) := \sum_{k \geq 0} e_k t^k = \prod_{i=1}^{n}(1 + x_i t).$$

The generating function $P(t) := \sum_{k \geq 1} p_k \frac{t^k}{k}$ of the *power sums* p_k is characterized by

$$H(t) = \exp\big(P(t)\big) = E(-t)^{-1}.$$

It follows that

$$h_d = \sum_{\mu \vdash d} \frac{1}{z_\mu} p_\mu \quad \text{and} \quad e_d = \sum_{\mu \vdash d} \frac{(-1)^{d-\ell(\mu)}}{z_\mu} p_\mu.$$

with $z_\mu := 1^{d_1} d_1! \, 2^{d_2} d_2! \cdots r^{d_r} d_r!$. The *involution* ω is such that $\omega(s_\lambda) = s_{\lambda'}$, thus $\omega(h_\lambda) = e_\lambda$ and $\omega(p_k) = (-1)^{k-1} p_k$. We start with $p_k[A + B] = p_k[A] + p_k[B]$, and $p_k[AB] = p_k[A] p_k[B]$, and we define *plethysm* to be linear and multiplicative: $(f + g)[A] = f[A] + g[A]$, $(f \cdot g)[A] = f[A] g[A]$. Then

$$f[-\mathbf{z}] = (-1)^d \omega\big(f(\mathbf{z})\big) \quad \text{and} \quad s_\lambda[-\mathbf{z}] = (-1)^n s_{\lambda'}(\mathbf{z}).$$

One can also check that

$$h_r(1, q, \ldots, q^{s-1}) = \begin{bmatrix} r + s - 1 \\ r \end{bmatrix}_q,$$

and more generally

$$s_\mu(1, q, q^2, \ldots) = \frac{q^{n(\mu)}}{\prod_{c \in \mu}(1 - q^{h(c)})}.$$

The *Vandermonde determinant* is

$$\Delta_n(\mathbf{x}) = \prod_{i < j}(x_i - x_j),$$

and *Schur functions* appear as $s_\lambda(\mathbf{x}) := \Delta_{\lambda + \delta_n}(\mathbf{x}) / \Delta_n(\mathbf{x})$. One also has, for *skew shapes*,

$$s_{\lambda/\mu}(\mathbf{x}) := \sum_\tau \mathbf{x}_\tau, \quad \text{with} \quad \mathbf{x}_\tau := \prod_{c \in \lambda} x_{\tau(c)},$$

where the sum is over semi-standard tableaux of shape λ/μ. In particular, this implies that

$$s_\lambda(\mathbf{x} + \mathbf{y}) = \sum_{\mu \subseteq \lambda} s_\mu(\mathbf{x}) s_{\lambda/\mu}(\mathbf{y}) \quad \text{and} \quad h_n(\mathbf{x} + \mathbf{y}) = \sum_{k=0}^n h_k(\mathbf{x}) h_{n-k}(\mathbf{y}).$$

The *Cauchy kernel* is

$$\Omega[\mathbf{xy}] := \prod_{i,j \geq 1} \frac{1}{1 - x_i y_j} = \sum_\lambda u_\lambda(\mathbf{x}) v_\lambda(\mathbf{y}),$$

where $\{u_\lambda\}_\lambda$ and $\{v_\mu\}_\mu$ are any *dual pair* of bases, i.e., $\langle u_\lambda, v_\mu \rangle = \delta_{\lambda\mu}$. The Schur basis is self dual, the h_μ are dual to the m_μ, and the dual basis of p_μ is p_μ / z_μ. The *Kronecker product* is the associative bilinear product such that $p_\lambda * p_\mu := \delta_{\lambda\mu} z_\lambda p_\lambda$. Then $H(1) = \sum_n h_n$ serves as identity, and multiplication by $E(1) = \sum_n e_n$ is equivalent to applying ω. The *Jacobi–Trudi* formulas are

$$s_{\mu/\lambda} = \det(h_{\mu_i - \lambda_j + j - i})_{1 \leq i,j \leq n} \quad \text{and} \quad s_{\mu'/\lambda'} = \det(e_{\mu_i - \lambda_j + j - i})_{1 \leq i,j \leq n}.$$

The *Littlewood–Richardson* rule describes the coefficients $c^\theta_{\lambda\mu}$ such that

$$s_\lambda s_\mu = \sum_{\theta \vdash |\lambda| + |\mu|} c^\theta_{\lambda\mu} s_\theta, \quad \text{or equivalently} \quad s_{\theta/\lambda} = \sum_\mu c^\theta_{\lambda\mu} s_\mu.$$

Particular cases are the *Pieri formulas*

$$h_k s_\mu = \sum_\theta s_\theta \quad \text{and} \quad e_k s_\mu = \sum_\theta s_\theta,$$

with θ/μ a k-celled *horizontal strip* (resp. *vertical strip*).

In terms of quasisymmetric functions, the Schur functions can be expressed as

$$s_{\lambda/\mu} = \sum_\tau Q_{\mathrm{co}(\tau)},$$

where $\mathrm{co}(\tau)$ is the composition encoding the reading descent set of τ. It follows that

$$s_{\lambda/\mu}[1/(1-q)] = \frac{\sum_\tau q^{\mathrm{rmaj}(\tau)}}{(1-q)(1-q^2)\cdots(1-q^n)}.$$

A.4   Integral Form Macdonald Functions

The *integral form Macdonald functions* are denoted by $H_\mu(\mathbf{z}; q, t)$. They expand in the Schur function basis as

$$H_\mu(\mathbf{z}; q, t) = \sum_{\lambda \vdash n} K_{\lambda,\mu}(q, t) s_\lambda(\mathbf{z}).$$

Considering the alternate scalar product defined by

$$\langle p_\lambda, p_\mu \rangle_* := \delta_{\lambda\mu}(-1)^{|\lambda|-\ell(\mu)} z_\lambda p_\lambda[(1-t)(1-q)],$$

we have $\langle H_\lambda, H_\mu \rangle_* = \delta_{\lambda\mu} \varepsilon_\mu \varepsilon'_\mu$. The H_μ are also characterized by the equations

$$\langle s_\lambda, H_\mu[\mathbf{z}(1-q); q, t] \rangle = 0, \quad \lambda \not\trianglerighteq \mu,$$
$$\langle s_\lambda, H_\mu[\mathbf{z}(1-t); q, t] \rangle = 0, \quad \lambda \not\trianglerighteq \mu',$$
$$\langle s_n, H_\mu(\mathbf{z}; q, t) \rangle = 1.$$

In combinatorial terms they appear as

$$H_\mu(\mathbf{z}; q, t) = \sum_{\tau:\, \mu \to \mathbb{N}} t^{\mathrm{maj}(\tau)} q^{\ell(\tau)} \mathbf{z}_\tau.$$

One has the specializations

$$H_\mu(\mathbf{z};0,0) = s_n(\mathbf{z}), \qquad H_\mu(\mathbf{z};0,1) = h_\mu(\mathbf{z}), \qquad H_\mu(\mathbf{z};1,1) = s_1(\mathbf{z})^n,$$
$$K_{\lambda,\mu}(0,0) = \delta_{\lambda,n}, \qquad K_{\lambda,\mu}(0,1) = K_{\lambda,\mu}, \qquad K_{\lambda,\mu}(1,1) = f^\lambda,$$

as well as

$$H_\mu[1-u;q,t] = \prod_{(i,j)\in\mu} (1 - uq^i t^j),$$

and the symmetries

$$H_{\mu'}(\mathbf{z};q,t) = H_\mu(\mathbf{z};t,q), \qquad T_\mu \omega H_\mu[\mathbf{z};q^{-1},t^{-1}] = H_\mu[\mathbf{z};q,t],$$
$$K_{\lambda,\mu}(q,t) = K_{\lambda,\mu'}(t,q), \qquad\qquad K_{\lambda,\mu}(q,t) = T_\mu K_{\lambda',\mu}(q^{-1},t^{-1}).$$

Important operators ∇ and $\mathbf{D}_k$ are defined by

$$\nabla H_\mu := T_\mu H_\mu \quad \text{and} \quad \sum_{k\in\mathbb{Z}} \mathbf{D}_k f(\mathbf{z}) y^k := f[\mathbf{z} + \theta/y]E[-y],$$

with $\theta = (1-t)(1-q)$. They satisfy the operator identities

$$[\mathbf{D}_m, s_1] = \theta \mathbf{D}_{m+1},$$
$$\nabla s_1 \nabla^{-1} = -\mathbf{D}_1,$$
$$\nabla^{-1} s_1^\perp \nabla = \frac{1}{\theta} \mathbf{D}_{-1},$$

where s_1 is understood to be the operator of multiplication by $s_1(\mathbf{z})$. We also have $\nabla^{-1} = \omega^* \nabla \omega^*$. One can characterize the H_μ by the fact that $\mathbf{D}_0 H_\mu = \lambda_\mu H_\mu$. Some other general identities concerning ∇ are

$$\nabla(e_{n-1}) = (\theta)^{-n-1} s_1^\perp \nabla(p_n),$$
$$\nabla(h_n)[1-u;q,t] = -u(-qt)^{n-1}\nabla(e_{n-1})[1-u;q,t].$$

Specializing ∇ at $t = 1$, we get a multiplicative operator such that

$$\nabla_{t=1}(e_n) = \sum_{\mu\vdash n} q^{n(\mu')} h_\mu\left[\frac{\mathbf{z}}{1-q}\right] f_\mu[1-q],$$

and the q-Catalan polynomials appear as $\langle \nabla_{t=1}(e_n), e_n \rangle = \mathcal{C}_n(q)$. Taking instead $t = 1/q$, we get

$$q^{\binom{n}{2}} \nabla_{t=1/q}(e_n) = \sum_{\mu\vdash n} q^{\binom{n}{2}+n(\mu')-n(\mu)} s_\mu\left[\frac{\mathbf{z}}{1-q}\right] s_{\mu'}[1-q],$$

and the other q-Catalan polynomials are obtained as $q^{\binom{n}{2}}\langle\nabla_{t=1/q}(e_n),e_n\rangle = \mathbf{C}_n(q)$. Still with this specialization, we also have

$$q^{\binom{n}{2}}\langle\nabla_{t=1/q}(e_n),p_1^n/n!\rangle = [n+1]_q^{n-1},$$

$$[n+1]_q s^\perp_{(n)}\nabla_{t=1/q}(e_n) = \begin{bmatrix} n+1 \\ k \end{bmatrix}_q q^{\binom{k}{2}-\binom{n}{2}}e_{n-k}\big[\mathbf{z}[n+k]_q\big].$$

A.5   Some Specific Values

The q,t-Catalan Polynomials for Small Values of n

$C_1(q,t) = 1,$

$C_2(q,t) = q + t,$

$C_3(q,t) = q^3 + q^2 t + qt^2 + t^3 + qt,$

$C_4(q,t) = q^6 + q^5 t + q^4 t^2 + q^3 t^3 + q^2 t^4 + qt^5 + t^6 + q^4 t + q^3 t^2 + q^2 t^3$
$\qquad\qquad + qt^4 + q^3 t + q^2 t^2 + qt^3,$

$C_5(q,t) = q^{10} + q^9 t + q^8 t^2 + q^7 t^3 + q^6 t^4 + q^5 t^5 + q^4 t^6 + q^3 t^7 + q^2 t^8$
$\qquad\qquad + qt^9 + t^{10} + q^8 t + q^7 t^2 + q^6 t^3 + q^5 t^4 + q^4 t^5 + q^3 t^6 + q^2 t^7$
$\qquad\qquad + qt^8 + q^7 t + 2q^6 t^2 + 2q^5 t^3 + 2q^4 t^4 + 2q^3 t^5 + 2q^2 t^6 + qt^7$
$\qquad\qquad + q^6 t + q^5 t^2 + 2q^4 t^3 + 2q^3 t^4 + q^2 t^5 + qt^6 + q^4 t^2 + q^3 t^3 + q^2 t^4,$

$C_6(q,t) = q^{15} + q^{14} t + q^{13} t^2 + q^{12} t^3 + q^{11} t^4 + q^{10} t^5 + q^9 t^6 + q^8 t^7 + q^7 t^8$
$\qquad\qquad + q^6 t^9 + q^5 t^{10} + q^4 t^{11} + q^3 t^{12} + q^2 t^{13} + qt^{14} + t^{15} + q^{13} t$
$\qquad\qquad + q^{12} t^2 + q^{11} t^3 + q^{10} t^4 + q^9 t^5 + q^8 t^6 + q^7 t^7 + q^6 t^8 + q^5 t^9$
$\qquad\qquad + q^4 t^{10} + q^3 t^{11} + q^2 t^{12} + qt^{13} + q^{12} t + 2q^{11} t^2 + 2q^{10} t^3 + 2q^9 t^4$
$\qquad\qquad + 2q^8 t^5 + 2q^7 t^6 + 2q^6 t^7 + 2q^5 t^8 + 2q^4 t^9 + 2q^3 t^{10} + 2q^2 t^{11}$
$\qquad\qquad + qt^{12} + q^{11} t + 2q^{10} t^2 + 3q^9 t^3 + 3q^8 t^4 + 3q^7 t^5 + 3q^6 t^6 + 3q^5 t^7$
$\qquad\qquad + 3q^4 t^8 + 3q^3 t^9 + 2q^2 t^{10} + qt^{11} + q^{10} t + 2q^9 t^2 + 3q^8 t^3 + 4q^7 t^4$
$\qquad\qquad + 4q^6 t^5 + 4q^5 t^6 + 4q^4 t^7 + 3q^3 t^8 + 2q^2 t^9 + qt^{10} + q^8 t^2 + 2q^7 t^3$
$\qquad\qquad + 3q^6 t^4 + 3q^5 t^5 + 3q^4 t^6 + 2q^3 t^7 + q^2 t^8 + q^7 t^2 + 2q^6 t^3 + 2q^5 t^4$
$\qquad\qquad + 2q^4 t^5 + 2q^3 t^6 + q^2 t^7 + q^4 t^4.$

Integral Form Macdonald Functions $H_\mu = H_\mu[x; q, t]$ for Partitions μ of n, $3 \leq n \leq 5$

$$H_3 = s_3 + (q^2 + q)s_{21} + q^3 s_{111},$$

$$H_{21} = s_3 + (q + t)s_{21} + qt s_{111},$$

$$H_{111} = s_3 + (t^2 + t)s_{21} + t^3 s_{111},$$

$$H_4 = s_4 + (q^3 + q^2 + q)s_{31} + (q^4 + q^2)s_{22} + (q^2 + q + 1)s_{211}q^3$$
$$+ q^6 s_{1111},$$

$$H_{31} = s_4 + (q^2 + q + t)s_{31} + (q^2 + qt)s_{22} + q(q^2 + qt + t)s_{211}$$
$$+ q^3 t s_{1111},$$

$$H_{22} = s_4 + (qt + q + t)s_{31} + (q^2 + t^2)s_{22} + qt(q + t + 1)s_{211}$$
$$+ q^2 t^2 s_{1111},$$

$$H_{211} = s_4 + (t^2 + q + t)s_{31} + (qt + t^2)s_{22} + t(qt + t^2 + q)s_{211} + qt^3 s_{1111},$$

$$H_{1111} = s_4 + (t^3 + t^2 + t)s_{31} + (t^4 + t^2)s_{22} + t^3(t^2 + t + 1)s_{211} + t^6 s_{1111},$$

$$H_5 = s_5 + (q^4 + q^3 + q^2 + q)s_{41} + (q^6 + q^5 + q^4 + q^3 + q^2)s_{32}$$
$$+ (q^7 + q^6 + 2q^5 + q^4 + q^3)s_{311} + q^4(q^4 + q^3 + q^2 + q + 1)s_{221}$$
$$+ q^6(q^3 + q^2 + q + 1)s_{2111} + q^{10} s_{11111},$$

$$H_{41} = s_5 + (q^3 + q^2 + q + t)s_{41} + (q^4 + q^3 + q^2 t + q^2 + qt)s_{32}$$
$$+ (q^5 + q^4 + q^3 t + q^3 + q^2 t + qt)s_{311}$$
$$+ q^2(q^3 + q^2 t + q^2 + qt + t)s_{221}$$
$$+ (q^3 + q^2 t + qt + t)s_{2111}q^3 + s_{11111}q^6 t,$$

$$H_{32} = s_5 + (q^2 + qt + q + t)s_{41} + (q^3 + q^2 t + q^2 + qt + t^2)s_{32}$$
$$+ (q^3 t + q^3 + 2q^2 t + qt^2 + qt)s_{311}$$
$$+ q(q^3 + q^2 t + qt^2 + qt + t^2)s_{221}$$
$$+ q^2 t(q^2 + qt + q + t)s_{2111} + q^4 t^2 s_{11111},$$

$$H_{311} = s_5 + (q^2 + t^2 + q + t)s_{41} + (q^2 t + qt^2 + q^2 + qt + t^2)s_{32}$$
$$+ (q^2 t^2 + q^3 + q^2 t + qt^2 + t^3 + qt)s_{311}$$
$$+ qt(q^2 + qt + t^2 + q + t)s_{221}$$
$$+ (q^2 t + qt^2 + q^2 + t^2)s_{2111}qt + s_{11111}q^3 t^3,$$

$$H_{221} = s_5 + (qt + t^2 + q + t)s_{41} + (qt^2 + t^3 + q^2 + qt + t^2)s_{32}$$
$$+ (qt^3 + q^2 t + 2qt^2 + t^3 + qt)s_{311}$$

$$+ t(q^2t + qt^2 + t^3 + q^2 + qt)s_{221}$$
$$+ qt^2(qt + t^2 + q + t)s_{2111} + q^2t^4 s_{11111},$$
$$H_{2111} = s_5 + (t^3 + t^2 + q + t)s_{41} + (t^4 + qt^2 + t^3 + qt + t^2)s_{32}$$
$$+ (t^5 + qt^3 + t^4 + qt^2 + t^3 + qt)s_{311}$$
$$+ t^2(qt^2 + t^3 + qt + t^2 + q)s_{221}$$
$$+ t^3(qt^2 + t^3 + qt + q)s_{2111} + qt^6 s_{11111},$$
$$H_{11111} = s_5 + (t^4 + t^3 + t^2 + t)s_{41} + (t^6 + t^5 + t^4 + t^3 + t^2)s_{32}$$
$$+ (t^7 + t^6 + 2t^5 + t^4 + t^3)s_{311} + t^4(t^4 + t^3 + t^2 + t + 1)s_{221}$$
$$+ t^6(t^3 + t^2 + t + 1)s_{2111} + t^{10}s_{11111}.$$

Bigraded Frobenius Characteristic of Diagonal Harmonics $\nabla(e_n) = \nabla(e_n)[x; q, t]$

$$\nabla(e_1) = s_1,$$
$$\nabla(e_2) = s_2 + (q + t)s_{11},$$
$$\nabla(e_3) = s_3 + (q^2 + qt + t^2 + q + t)s_{21} + (q^3 + q^2t + qt^2 + t^3 + qt)s_{111},$$
$$\nabla(e_4) = s_4 + (q^3 + q^2t + qt^2 + t^3 + q^2 + qt + t^2 + q + t)s_{31}$$
$$+ (q^4 + q^3t + q^2t^2 + qt^3 + t^4 + q^2t + qt^2 + q^2 + qt + t^2)s_{22}$$
$$+ (q^5 + q^4t + q^3t^2 + q^2t^3 + qt^4 + t^5 + q^4 + 2q^3t + 2q^2t^2$$
$$+ 2qt^3 + t^4 + q^3 + 2q^2t + 2qt^2 + t^3 + qt)s_{211}$$
$$+ (q^6 + q^5t + q^4t^2 + q^3t^3 + q^2t^4 + qt^5 + t^6 + q^4t + q^3t^2$$
$$+ q^2t^3 + qt^4 + q^3t + q^2t^2 + qt^3)s_{1111},$$
$$\nabla(e_5) = s_5 + (q^4 + q^3t + q^2t^2 + qt^3 + t^4 + q^3 + q^2t$$
$$+ qt^2 + t^3 + q^2 + qt + t^2 + q + t)s_{41}$$
$$+ (q^6 + q^5t + q^4t^2 + q^3t^3 + q^2t^4 + qt^5 + t^6 + q^5 + 2q^4t + 2q^3t^2$$
$$+ 2q^2t^3 + 2qt^4 + t^5 + q^4 + 2q^3t + 3q^2t^2 + 2qt^3 + t^4 + q^3$$
$$+ 2q^2t + 2qt^2 + t^3 + q^2 + qt + t^2)s_{32}$$
$$+ (q^7 + q^6t + q^5t^2 + q^4t^3 + q^3t^4 + q^2t^5 + qt^6 + t^7 + q^6 + 2q^5t$$
$$+ 2q^4t^2 + 2q^3t^3 + 2q^2t^4 + 2qt^5 + t^6 + 2q^5 + 3q^4t + 4q^3t^2$$
$$+ 4q^2t^3 + 3qt^4 + 2t^5 + q^4 + 3q^3t + 3q^2t^2 + 3qt^3 + t^4 + q^3$$
$$+ 2q^2t + 2qt^2 + t^3 + qt)s_{311}$$

$$
\begin{aligned}
+ (q^8 &+ q^7t + q^6t^2 + q^5t^3 + q^4t^4 + q^3t^5 + q^2t^6 + qt^7 + t^8 + q^7 \\
&+ 2q^6t + 2q^5t^2 + 2q^4t^3 + 2q^3t^4 + 2q^2t^5 + 2qt^6 + t^7 + q^6 \\
&+ 3q^5t + 4q^4t^2 + 4q^3t^3 + 4q^2t^4 + 3qt^5 + t^6 + q^5 \\
&+ 3q^4t4q^3t^2 + 4q^2t^3 + 3qt^4 + t^5 + q^4 + 2q^3t + 3q^2t^2 + 2qt^3 \\
&+ t^4 + q^2t + qt^2)s_{221} \\
+ (q^9 &+ q^8t + q^7t^2 + q^6t^3 + q^5t^4 + q^4t^5 + q^3t^6 + q^2t^7 + qt^8 + t^9 \\
&+ q^8 + 2q^7t + 2q^6t^2 + 2q^5t^3 + 2q^4t^4 + 2q^3t^5 + 2q^2t^6 + 2qt^7 \\
&+ t^8 + q^7 + 3q^6t + 4q^5t^2 + 4q^4t^3 + 4q^3t^4 + 4q^2t^5 + 3qt^6 \\
&+ t^7 + q^6 + 3q^5t + 4q^4t^2 + 5q^3t^3 + 4q^2t^4 + 3qt^5 + t^6 + 2q^4t \\
&+ 3q^3t^2 + 3q^2t^3 + 2qt^4 + q^3t + q^2t^2 + qt^3)s_{2111} \\
+ (q^{10} &+ q^9t + q^8t^2 + q^7t^3 + q^6t^4 + q^5t^5 + q^4t^6 + q^3t^7 + q^2t^8 \\
&+ qt^9 + t^{10} + q^8t + q^7t^2 + q^6t^3 + q^5t^4 + q^4t^5 + q^3t^6 + q^2t^7 \\
&+ qt^8 + q^7t + 2q^6t^2 + 2q^5t^3 + 2q^4t^4 + 2q^3t^5 + 2q^2t^6 + qt^7 \\
&+ q^6t + q^5t^2 + 2q^4t^3 + 2q^3t^4 + q^2t^5 + qt^6 + q^4t^2 \\
&+ q^3t^3 + q^2t^4)s_{11111}.
\end{aligned}
$$

Bibliography

[Artin 44] E. Artin. *Galois Theory*, Second edition, Notre Dame Mathematical Lectures, no. 2. Notre Dame, Ind.: University of Notre Dame, 1944.

[Atiyah and Macdonald 69] M. F. Atiyah and I. G. Macdonald. *Introduction to commutative algebra*. Addison-Wesley Publishing Co., Reading, Mass.-London-Don Mills, Ont., 1969.

[Aval 02] J.-C. Aval. "On certain spaces of lattice diagram determinants". *Discrete Math.* 256:3 (2002), 557–575. LaCIM 2000 Conference on Combinatorics, Computer Science and Applications (Montreal, QC).

[Aval et al. 02] J.-C. Aval, F. Bergeron, and N. Bergeron. "Spaces of lattice diagram polynomials in one set of variables". *Adv. in Appl. Math.* 28:3-4 (2002), 343–359. Special issue in memory of Rodica Simion.

[Aval et al. 04] J.-C. Aval, F. Bergeron, and N. Bergeron. "Ideals of quasi-symmetric functions and super-covariant polynomials for $\mathfrak{S}_n$". *Adv. Math.* 181:2 (2004), 353–367.

[Aval et al. 07] J.-C. Aval, F. Bergeron, and N. Bergeron. "Diagonal Temperley-Lieb invariants and harmonics". *Sém. Lothar. Combin.* 54A (2005/07), Art. B54Aq, 19 pp. (electronic).

[Baker and Forrester 97] T. H. Baker and P. J. Forrester. "The Calogero-Sutherland model and polynomials with prescribed symmetry". *Nuclear Phys. B* 492:3 (1997), 682–716.

[Bergeron 90] F. Bergeron. "Combinatoire des polynômes orthogonaux classiques: une approche unifiée". *European J. Combin.* 11:5 (1990), 393–401.

[Bergeron and Bergeron 05] F. Bergeron and N. Bergeron. "Is r-Qsym Free over Sym?" FPSAC: Special session in honour of Adriano Garsia's 75th birthday. Taormina, Sicily, 2005.

[Bergeron and Biagioli 06] F. Bergeron and R. Biagioli. "Tensorial square of the hyperoctahedral group coinvariant space". *Electron. J. Combin.* 13:1 (2006), Research Paper 38, 32 pp. (electronic).

[Bergeron and Garsia 99] F. Bergeron and A. M. Garsia. "Science fiction and Macdonald's polynomials". In *Algebraic methods and q-special functions (Montréal, QC, 1996)*, CRM Proc. Lecture Notes, 22, pp. 1–52. Providence, RI: Amer. Math. Soc., 1999.

[Bergeron and Hamel 00] F. Bergeron and S. Hamel. "Intersection of modules related to Macdonald's polynomials". *Discrete Math.* 217:1-3 (2000), 51–64. Formal power series and algebraic combinatorics (Vienna, 1997).

[Bergeron and Lamontagne 07] F. Bergeron and F. Lamontagne. "Decomposition of the diagonal action of S_n on the coinvariant space of $S_n \times S_n$". *Sém. Lothar. Combin.* 52 (2004/07), Art. B52e, 24 pp. (electronic).

[Bergeron et al. 98] F. Bergeron, G. Labelle, and P. Leroux. *Combinatorial species and tree-like structures*, Encyclopedia of Mathematics and Its Applications, 67. Cambridge: Cambridge University Press, 1998. Translated from the 1994 French original by Margaret Readdy, with a foreword by Gian-Carlo Rota.

[Bergeron et al. 99a] F. Bergeron, A. M. Garsia, M. Haiman, and G. Tesler. "Identities and positivity conjectures for some remarkable operators in the theory of symmetric functions". *Methods Appl. Anal.* 6:3 (1999), 363–420. Dedicated to Richard A. Askey on the occasion of his 65th birthday, Part III.

[Bergeron et al. 99b] F. Bergeron, N. Bergeron, A. M. Garsia, M. Haiman, and G. Tesler. "Lattice diagram polynomials and extended Pieri rules". *Adv. Math.* 142:2 (1999), 244–334.

[Bergeron et al. 00] F. Bergeron, A. Garsia, and G. Tesler. "Multiple left regular representations generated by alternants". *J. Combin. Theory Ser. A* 91:1-2 (2000), 49–83. In memory of Gian-Carlo Rota.

[Blessenohl and Schocker 05] D. Blessenohl and M. Schocker. *Noncommutative Character Theory of the Symmetric Group*. London: Imperial College Press, 2005.

[Borel 53] A. Borel. "Sur la cohomologie des espaces fibrés principaux et des espaces homogènes de groupes de Lie compacts". *Ann. of Math. (2)* 57 (1953), 115–207.

[Can and Loehr 06] M. Can and N. Loehr. "A proof of the q,t-square conjecture". *J. Combin. Theory Ser. A* 113:7 (2006), 1419–1434.

[Cox et al. 92] D. Cox, J. Little, and D. O'Shea. *Ideals, varieties, and algorithms*, Undergraduate Texts in Mathematics. New York: Springer-Verlag, 1992. An introduction to computational algebraic geometry and commutative algebra.

[De Concini and Procesi 81] C. De Concini and C. Procesi. "Symmetric functions, conjugacy classes and the flag variety". *Invent. Math.* 64:2 (1981), 203–219.

[Etingof and Strickland 03] P. Etingof and E. Strickland. "Lectures on quasi-invariants of Coxeter groups and the Cherednik algebra". *Enseign. Math. (2)* 49:1-2 (2003), 35–65.

[Felder and Veselov 03] G. Felder and A. P. Veselov. "Action of Coxeter groups on m-harmonic polynomials and Knizhnik-Zamolodchikov equations". *Mosc. Math. J.* 3:4 (2003), 1269–1291.

[Foata 84] D. Foata. "Combinatoire des identités sur les polynômes orthogonaux". In *Proceedings of the International Congress of Mathematicians, Vol. 1, 2 (Warsaw, 1983)*, pp. 1541–1553. Warsaw: PWN, 1984.

[Foata and Leroux 83] D. Foata and P. Leroux. "Polynômes de Jacobi, interprétation combinatoire et fonction génératrice". *Proc. Amer. Math. Soc.* 87:1 (1983), 47–53.

[Frame et al. 54] J. S. Frame, G. de B. Robinson, and R. M. Thrall. "The hook graphs of the symmetric groups". *Canadian J. Math.* 6 (1954), 316–324.

[Franklin 1881] F. Franklin. "Sur le développement du produit infini $(1-x)(1-x^2)(1-x^3)\cdots$". *C. R. Acad. Sci. Paris* 92 (1881), 448–450.

[Fulton 97] W. Fulton. *Young tableaux*, London Mathematical Society Student Texts, 35. Cambridge: Cambridge University Press, 1997. With applications to representation theory and geometry.

[Fulton and Harris 91] W. Fulton and J. Harris. *Representation theory*, Graduate Texts in Mathematics, 129. New York: Springer-Verlag, 1991. A first course, Readings in Mathematics.

[Garsia 81] A. M. Garsia. "A q-analogue of the Lagrange inversion formula". *Houston J. Math.* 7:2 (1981), 205–237.

[Garsia and Haglund 02] A. M. Garsia and J. Haglund. "A proof of the q, t-Catalan positivity conjecture". *Discrete Math.* 256:3 (2002), 677–717. LaCIM 2000 Conference on Combinatorics, Computer Science and Applications (Montreal, QC).

[Garsia and Haiman 93] A. M. Garsia and M. Haiman. "A graded representation model for Macdonald's polynomials". *Proc. Nat. Acad. Sci. U.S.A.* 90:8 (1993), 3607–3610.

[Garsia and Haiman 96a] A. M. Garsia and M. Haiman. "A remarkable q, t-Catalan sequence and q-Lagrange inversion". *J. Algebraic Combin.* 5:3 (1996), 191–244.

[Garsia and Haiman 96b] A. M. Garsia and M. Haiman. "Some natural bigraded S_n-modules and q, t-Kostka coefficients". *Electron. J. Combin.* 3:2 (1996), Research Paper 24, approx. 60 pp. (electronic). The Foata Festschrift.

[Garsia and Haiman 08] A. M. Garsia and M. Haiman. "Orbit Harmonics and Graded Representations". Publications du LaCIM, Université du Québec à Montréal. Manuscript, 2008.

[Garsia and Remmel 05] A. Garsia and J. B. Remmel. "Breakthroughs in the theory of Macdonald polynomials". *Proc. Natl. Acad. Sci. USA* 102:11 (2005), 3891–3894 (electronic).

[Garsia and Tesler 96] A. M. Garsia and G. Tesler. "Plethystic formulas for Macdonald q, t-Kostka coefficients". *Adv. Math.* 123:2 (1996), 144–222.

[Garsia and Wallach 03] A. M. Garsia and N. Wallach. "Qsym over Sym is free". *J. Combin. Theory Ser. A* 104:2 (2003), 217–263.

[Gessel and Viennot 85] I. Gessel and G. Viennot. "Binomial determinants, paths, and hook length formulae". *Adv. in Math.* 58:3 (1985), 300–321.

[Goodman and Wallach 98] R. Goodman and N. R. Wallach. *Representations and invariants of the classical groups*, Encyclopedia of Mathematics and Its Applications, 68. Cambridge: Cambridge University Press, 1998.

[Haglund 08] J. Haglund. *The q,t-Catalan numbers and the space of diagonal harmonics*, University Lecture Series, 41. Providence, RI: American Mathematical Society, 2008. With an appendix on the combinatorics of Macdonald polynomials.

[Haglund et al. 05a] J. Haglund, M. Haiman, and N. Loehr. "A combinatorial formula for Macdonald polynomials". *J. Amer. Math. Soc.* 18:3 (2005), 735–761 (electronic).

[Haglund et al. 05b] J. Haglund, M. Haiman, N. Loehr, J. B. Remmel, and A. Ulyanov. "A combinatorial formula for the character of the diagonal coinvariants". *Duke Math. J.* 126:2 (2005), 195–232.

[Haiman 01] M. Haiman. "Hilbert schemes, polygraphs and the Macdonald positivity conjecture". *J. Amer. Math. Soc.* 14:4 (2001), 941–1006 (electronic).

[Hivert 00] F. Hivert. "Hecke algebras, difference operators, and quasi-symmetric functions". *Adv. Math.* 155:2 (2000), 181–238.

[Hivert 08] F. Hivert. "Local Action of the Symmetric Group and Generalizations of Quasi-Symmetric Functions". Preprint, 2008.

[Hivert and Thiéry 04] F. Hivert and N. M. Thiéry. "Deformation of symmetric functions and the rational Steenrod algebra". In *Invariant theory in all characteristics*, CRM Proc. Lecture Notes, 35, pp. 91–125. Providence, RI: Amer. Math. Soc., 2004.

[Hoffman 00] M. E. Hoffman. "Quasi-shuffle products". *J. Algebraic Combin.* 11:1 (2000), 49–68.

[Humphreys 90] J. E. Humphreys. *Reflection groups and Coxeter groups*, Cambridge Studies in Advanced Mathematics, 29. Cambridge: Cambridge University Press, 1990.

[Jacobi 41] C. Jacobi. "De functionibus alternantibus earumque divisione per productum e differentiis elementorum conflatum". *J. Reine Angew. Math. (Crelle)* 22 (1841), 360–371. Also in Mathematische Werke Volume 3, Chelsea, 1969, 439–452.

[Joyal 81] A. Joyal. "Une théorie combinatoire des séries formelles". *Adv. in Math.* 42:1 (1981), 1–82.

[Joyal 86] A. Joyal. "Foncteurs analytiques et espèces de structures". In *Combinatoire énumérative (Montreal, Que., 1985/Quebec, Que., 1985)*, Lecture Notes in Math., 1234, pp. 126–159. Berlin: Springer, 1986.

[Kac and Cheung 02] V. Kac and P. Cheung. *Quantum calculus*. Universitext, New York: Springer-Verlag, 2002.

[Kane 01] R. Kane. *Reflection groups and invariant theory*, CMS Books in Mathematics/Ouvrages de Mathématiques de la SMC, 5. New York: Springer-Verlag, 2001.

[Kirillov and Noumi 98] A. N. Kirillov and M. Noumi. "Affine Hecke algebras and raising operators for Macdonald polynomials". *Duke Math. J.* 93:1 (1998), 1–39.

[Knop 97] F. Knop. "Integrality of two variable Kostka functions". *J. Reine Angew. Math.* 482 (1997), 177–189.

[Knuth 70] D. E. Knuth. "Permutations, matrices, and generalized Young tableaux". *Pacific J. Math.* 34 (1970), 709–727.

[Knutson 73] D. Knutson. λ-*rings and the representation theory of the symmetric group*, Lecture Notes in Mathematics, 308. Berlin: Springer-Verlag, 1973.

[Labelle and Yeh 89] J. Labelle and Y. N. Yeh. "The combinatorics of Laguerre, Charlier, and Hermite polynomials". *Stud. Appl. Math.* 80:1 (1989), 25–36.

[Lapointe and Morse 08] L. Lapointe and J. Morse. "Quantum cohomology and the k-Schur basis". *Trans. Amer. Math. Soc.* 360:4 (2008), 2021–2040.

[Lapointe and Vinet 97] L. Lapointe and L. Vinet. "Rodrigues formulas for the Macdonald polynomials". *Adv. Math.* 130:2 (1997), 261–279.

[Lascoux and Schützenberger 78] A. Lascoux and M.-P. Schützenberger. "Sur une conjecture de H. O. Foulkes". *C. R. Acad. Sci. Paris Sér. A-B* 286:7 (1978), A323–A324.

[Lascoux et al. 97] A. Lascoux, B. Leclerc, and J.-Y. Thibon. "Ribbon tableaux, Hall-Littlewood functions, quantum affine algebras, and unipotent varieties". *J. Math. Phys.* 38:2 (1997), 1041–1068.

[Lenart 00] C. Lenart. "Lagrange inversion and Schur functions". *J. Algebraic Combin.* 11:1 (2000), 69–78.

[Leroux and Strehl 85] P. Leroux and V. Strehl. "Jacobi polynomials: combinatorics of the basic identities". *Discrete Math.* 57:1-2 (1985), 167–187.

[Lindström 73] B. Lindström. "On the vector representations of induced matroids". *Bull. London Math. Soc.* 5 (1973), 85–90.

[Littlewood 50] D. E. Littlewood. *The Theory of Group Characters*, Second edition. New York: Oxford University Press, 1950.

[Littlewood 56] D. E. Littlewood. "The Kronecker product of symmetric group representations". *J. London Math. Soc.* 31 (1956), 89–93.

[Lothaire 02] M. Lothaire. *Algebraic combinatorics on words*, Encyclopedia of Mathematics and Its Applications, 90. Cambridge: Cambridge University Press, 2002.

[Macdonald 88] I. G. Macdonald. *A New Class of Symmetric Functions*, pp. 131–171, Actes 20e Séminaire Lotharingien. Strasbourg: I.R.M.A., 1988.

[Macdonald 95] I. G. Macdonald. *Symmetric functions and Hall polynomials*, Second edition, Oxford Mathematical Monographs. New York: The Clarendon Press Oxford University Press, 1995. With contributions by A. Zelevinsky, Oxford Science Publications.

[Macdonald 03] I. G. Macdonald. *Affine Hecke algebras and orthogonal polynomials*, Cambridge Tracts in Mathematics, 157. Cambridge: Cambridge University Press, 2003.

[Malvenuto and Reutenauer 95] C. Malvenuto and C. Reutenauer. "Duality between quasi-symmetric functions and the Solomon descent algebra". *J. Algebra* 177:3 (1995), 967–982.

[Manivel 01] L. Manivel. *Symmetric functions, Schubert polynomials and degeneracy loci*, SMF/AMS Texts and Monographs, 6. Providence, RI: American Mathematical Society, 2001. Translated from the 1998 French original by John R. Swallow, Cours Spécialisés [Specialized Courses], 3.

[Robinson 38] G. de B. Robinson. "On the Representations of the Symmetric Group". *Amer. J. Math.* 60:3 (1938), 745–760.

[Roth 05] M. Roth. "Inverse systems and regular representations". *J. Pure Appl. Algebra* 199:1-3 (2005), 219–234.

[Sagan 91] B. E. Sagan. *The symmetric group*, The Wadsworth & Brooks/Cole Mathematics Series. Pacific Grove, CA: Wadsworth & Brooks/Cole Advanced Books & Software, 1991. Representations, combinatorial algorithms, and symmetric functions.

[Sahi 96] S. Sahi. "Interpolation, integrality, and a generalization of Macdonald's polynomials". *Internat. Math. Res. Notices* :10 (1996), 457–471.

[Schensted 61] C. Schensted. "Longest increasing and decreasing subsequences". *Canad. J. Math.* 13 (1961), 179–191.

[Schur 01] I. Schur. "Uber eine Klasse von Matrizen die sich einer gegeben Matrix zuordnen lassen". 1901, Inaugural-Dissertation, Berlin.

[Stanley 79] R. P. Stanley. "Invariants of finite groups and their applications to combinatorics". *Bull. Amer. Math. Soc. (N.S.)* 1:3 (1979), 475–511.

[Stanley 96] R. P. Stanley. *Combinatorics and commutative algebra*, Second edition, Progress in Mathematics, 41. Boston, MA: Birkhäuser Boston Inc., 1996.

[Stanley 97] R. P. Stanley. *Enumerative combinatorics, Vol. 1*, Cambridge Studies in Advanced Mathematics, 49. Cambridge: Cambridge University Press, 1997. With a foreword by Gian-Carlo Rota. Corrected reprint of the 1986 original.

[Steinberg 64] R. Steinberg. "Differential equations invariant under finite reflection groups". *Trans. Amer. Math. Soc.* 112 (1964), 392–400.

[Stembridge 09] J. Stembridge. "The SF Package". Software package. Available from World Wide Web (http://www.math.lsa.umich.edu/~jrs/maple.html#SF), 2009.

[Viennot 77] G. Viennot. "Une forme géométrique de la correspondance de Robinson-Schensted". In *Combinatoire et représentation du groupe symétrique (Actes Table Ronde CNRS, Univ. Louis-Pasteur Strasbourg, Strasbourg, 1976)*, pp. 29–58. Lecture Notes in Mathematics, 579. Berlin: Springer, 1977.

[Wood 98] R. M. W. Wood. "Problems in the Steenrod algebra". *Bull. London Math. Soc.* 30:5 (1998), 449–517.

Index

CMS Treatises in Mathematics

Other titles in the series:

Summa Summarum
Mogens Esrom Larsen
ISBN 978-1-56881-323-3

"The book is quite comprehensive and discusses a host of techniques from the classical ideas of Euler to the modern ideas of R. W. Gosper, Jr., H Wilf, and D. Zeilberger, of how to simplify finite sums that are likely to appear in the course of ones work. ... This work should prove to be an invaluable aid to students and researchers working in all areas of mathematics. The author's 'hope is to find this summa on your desk—just as Thomas's original was found on the altar!' and the reviewer agrees."

—*Mathematical Reviews*

Factorization: Unique and Otherwise
Steven H. Weintraub
ISBN 978-1-56881-241-0

"The concept of factorization, familiar in the ordinary system of whole numbers that can be written as a unique product of prime numbers, plays a central role in modern mathematics and its applications. This exposition of the classic theory leads the reader to an understanding of the current knowledge of the subject and its connections to other mathematical concepts You will learn that instead of unique factorization being the norm and non-unique factorization the exception, the situation is reversed!"

—*L'Enseignement Mathematique*

Books are available for order through A K Peters. You may find detailed ordering information on our website at www.akpeters.com.

Canadian Mathematical Society
Société mathématique du Canada
Ottawa, Ontario

A K Peters, Ltd.
Wellesley, Massachusetts